Springer Monographs in Mathematics

Springer
Berlin
Heidelberg
New York
Hong Kong
London
Milan
Paris
Tokyo

Springer
Berlin
Heidelberg
New York
Barcelona
Hong Kong
London
Milan
Paris
Singapore
Tokyo

Jean-Pierre Serre

Local Algebra

Translated from the French by CheeWhye Chin

Springer

Jean-Pierre Serre
Collège de France
3 rue d'Ulm
75005 Paris, France
e-mail: serre@dmi.ens.fr

Translator:
CheeWhye Chin
Princeton University
Department of Mathematics
Princeton, NJ 08544
USA
e-mail: cchin@princeton.edu

CIP data applied for

Die Deutsche Bibliothek – CIP-Einheitsaufnahme

Serre, Jean-Pierre:
Local algebra / Jean-Pierre Serre. Transl. from the French by CheeWhye Chin.- Berlin; Heidelberg;
New York; Barcelona; Hong Kong; London; Milan; Paris; Singapore; Tokyo: Springer, 2000
(Springer monographs in mathematics)
Einheitssacht.: Algèbre locale <engl.>

Mathematics Subject Classification (2000): 13xx

ISSN 1439-7382
ISBN 978-3-642-08590-1

Springer-Verlag is a company in the BertelsmannSpringer publishing group.
© Springer-Verlag Berlin Heidelberg 2000
Softcover reprint of the hardcover 1st edition 2000

Cover design: *Erich Kirchner, Heidelberg*

Printed on acid-free paper

Preface

The present book is an English translation of

Algèbre Locale — Multiplicités

published by Springer-Verlag as no. 11 of the Lecture Notes series.

The original text was based on a set of lectures, given at the Collège de France in 1957-1958, and written up by Pierre Gabriel. Its aim was to give a short account of Commutative Algebra, with emphasis on the following topics:

a) *Modules* (as opposed to *Rings*, which were thought to be the only subject of Commutative Algebra, before the emergence of sheaf theory in the 1950s);

b) *Homological methods*, à la Cartan-Eilenberg;

c) *Intersection multiplicities*, viewed as Euler-Poincaré characteristics.

The English translation, done with great care by CheeWhye Chin, differs from the original in the following aspects:
- The terminology has been brought up to date (e.g. "cohomological dimension" has been replaced by the now customary "depth").
- I have rewritten a few proofs and clarified (or so I hope) a few more.
- A section on graded algebras has been added (App. III to Chap. IV).
- New references have been given, especially to other books on Commutative Algebra: Bourbaki (whose Chap. X has now appeared, after a 40-year wait), Eisenbud, Matsumura, Roberts, ...

I hope that these changes will make the text easier to read, without changing its informal "Lecture Notes" character.

J-P. Serre,
Princeton, Fall 1999

Contents

x   Contents

Introduction

The intersection multiplicities of algebraic geometry are equal to some "Euler-Poincaré characteristics" constructed by means of the Tor functor of Cartan-Eilenberg. The main purpose of this course is to prove this result, and to apply it to the fundamental formulae of intersection theory.

It is necessary to first recall some basic results of local algebra: primary decomposition, Cohen-Seidenberg theorems, normalization of polynomial rings, Krull dimension, characteristic polynomials (in the sense of Hilbert-Samuel).

Homology comes next, when we consider the multiplicity $e_q(E, r)$ of an ideal of definition $\mathfrak{q} = (x_1, \ldots, x_r)$ of a local noetherian ring A with respect to a finitely generated A-module E. This multiplicity is defined as the coefficient of $n^r/r!$ in the polynomial-like function $n \mapsto \ell_A(E/\mathfrak{q}^n E)$ [here $\ell_A(F)$ is the length of an A-module F]. We prove in this case the following formula, which plays an essential role in the sequel:

$$e_q(E, r) = \sum_{i=0}^{r} (-1)^i \ell_A(H_i(\mathbf{x}, E)) \qquad (*)$$

where the $H_i(\mathbf{x}, E)$ denotes the homology modules of the Koszul complex constructed on E by means of $\mathbf{x} = (x_1, \ldots, x_r)$.

Moreover this complex can be used in other problems of local algebra, for example for the study of the depth of modules over a local ring and of the Cohen-Macaulay modules (those whose Krull dimension coincides with their depth), and also for showing that regular local rings are the only local rings whose homological dimension is finite.

Once formula $(*)$ is proved, one may study the Euler-Poincaré characteristic constructed by means of Tor. When one translates the geometric situation of intersections into the language of local algebra, one obtains a regular local ring A, of dimension n, and two finitely generated A-modules E and F over A, whose tensor product is of finite length over A (this means that the varieties corresponding to E and F intersect only at the given point). One is then led to conjecture the following statements:

(i) $\dim(E) + \dim(F) \leq n$ ("dimension formula").

(ii) $\chi_A(E, F) = \sum_{i=0}^{n}(-1)^i \ell_A(\mathrm{Tor}_i^A(E, F))$ is ≥ 0.

(iii) $\chi_A(E, F) = 0$ if and only if the inequality in (i) is strict.

Formula (∗) shows that the statements (i), (ii) and (iii) are true if $F = A/(x_1, \ldots, x_r)$, with $\dim(F) = n - r$. Thanks to a process, using completed tensor products, which is the algebraic analogue of "reduction to the diagonal", one can show that they are true when A has the same characteristic as its residue field, or when A is unramified. To go beyond that, one can use the structure theorems of complete local rings to prove (i) in the most general case. On the other hand, I have not succeeded in proving (ii) and (iii) without making assumptions about A, nor to give counter-examples. It seems that it is necessary to approach the question from a different angle, for example by directly defining (by a suitable asymptotic process) an integer ≥ 0 which one would subsequently show to be equal to $\chi_A(E, F)$.

Fortunately, the case of equal characteristic is sufficient for the applications to algebraic geometry (and also to analytic geometry). More specifically, let X be a non-singular variety, let V and W be two irreducible subvarieties of X, and suppose that $C = V \cap W$ is an irreducible subvariety of X, with:

$$\dim X + \dim C = \dim V + \dim W \qquad \text{("proper" intersection)}.$$

Let A, A_V, A_W be the local rings of X, V and W at C. If

$$i(V \cdot W, C; X)$$

denotes the multiplicity of the intersection of V and W at C (in the sense of Weil, Chevalley, Samuel), we have the formula:

$$i(V \cdot W, C; X) = \chi_A(A_V, A_W). \qquad (**)$$

This formula is proved by reduction to the diagonal, and the use of (∗). In fact, it is convenient to take (∗∗) as the definition of multiplicities. The properties of these multiplicities are then obtained in a natural way: commutativity follows from that for Tor; associativity follows from the two spectral sequences which expresses the associativity of Tor; the projection formula follows from the two spectral sequences connecting the direct images of a coherent sheave and Tor (these latter spectral sequences have other interesting applications, but they are not explored in the present course). In each case, one uses the well-known fact that Euler-Poincaré characteristics remain constant through a spectral sequence.

When one defines intersection multiplicities by means of the Tor-formula above, one is led to extend the theory beyond the strictly "non-singular" framework of Weil and Chevalley. For example, if $f : X \to Y$

is a morphism of a variety X into a non-singular variety Y, one can associate, to two cycles x and y of X and Y, a "product" $x \cdot_f y$ which corresponds to $x \cap f^{-1}(y)$ (of course, this product is only defined under certain dimension conditions). When f is the identity map, one recovers the standard product. The commutativity, associativity and projection formulae can be stated and proved for this new product.

Chapter I. Prime Ideals and Localization

This chapter summarizes standard results in commutative algebra. For more details, see [Bour], Chap. II, III, IV.

1. Notation and definitions

In what follows, all rings are commutative, with a unit element 1.

An ideal $\mathfrak{p}$ of a ring A is called **prime** if $A/\mathfrak{p}$ is a **domain**, i.e. can be embedded into a field; such an ideal is distinct from A.

An ideal $\mathfrak{m}$ of A is called **maximal** if it is distinct from A, and maximal among the ideals having this property; it amounts to the same as saying that $A/\mathfrak{m}$ is a field. Such an ideal is prime.

A ring A is called **semilocal** if the set of its maximal ideals is finite. It is called **local** if it has one and only one maximal ideal $\mathfrak{m}$; one then has $A - \mathfrak{m} = A^*$, where A^* denotes the multiplicative group of invertible elements of A.

2. Nakayama's lemma

Let $\mathfrak{r}$ be the **Jacobson radical** of A, i.e. the intersection of all maximal ideals of A. Then $x \in \mathfrak{r}$ if and only if $1 - xy$ is invertible for every $y \in A$.

Proposition 1. *Let M be a finitely generated A-module, and $\mathfrak{q}$ be an ideal of A contained in the radical $\mathfrak{r}$ of A. If $\mathfrak{q}M = M$, then $M = 0$.*

Indeed, if M is $\neq 0$, it has a quotient which is a simple module, hence is isomorphic to $A/\mathfrak{m}$, where $\mathfrak{m}$ is a maximal ideal of A; then $\mathfrak{m}M \neq M$, contrary to the fact that $\mathfrak{q} \subset \mathfrak{m}$.

Corollary 1.   *If N is a submodule of M such that $M = N + \mathfrak{q}M$, we have $M = N$.*

This follows from prop. 1, applied to M/N.

Corollary 2.   *If A is a local ring, and if M and N are two finitely generated A-modules, then:*

$$M \otimes_A N = 0 \iff (M = 0 \quad \text{or} \quad N = 0).$$

Let $\mathfrak{m}$ be the maximal ideal of A, and k the field $A/\mathfrak{m}$. Set

$$\overline{M} = M/\mathfrak{m}M \quad \text{and} \quad \overline{N} = N/\mathfrak{m}N.$$

If $M \otimes_A N$ is zero, so is $\overline{M} \otimes_k \overline{N}$; this implies $\overline{M} = 0$ or $\overline{N} = 0$, whence $M = 0$ or $N = 0$, according to prop. 1.

3. Localization   (cf. [Bour], Chap. II)

Let S be a subset of A closed under multiplication, and containing 1. If M is an A-module, the module $S^{-1}M$ (sometimes also written as M_S) is defined as the set of "fractions" m/s, $m \in M$, $s \in S$, two fractions m/s and m'/s' being identified if and only if there exists $s'' \in S$ such that $s''(s'm - sm') = 0$. This also applies to $M = A$, which defines $S^{-1}A$. We have natural maps

$$A \to S^{-1}A \quad \text{and} \quad M \to S^{-1}M$$

given by $a \mapsto a/1$ and $m \mapsto m/1$. The kernel of $M \to S^{-1}M$ is $\text{Ann}_M(S)$, i.e. the set of $m \in M$ such that there exists $s \in S$ with $sm = 0$.

The multiplication rule

$$a/s \cdot a'/s' = (aa')/(ss')$$

defines a ring structure on $S^{-1}A$. Likewise, the module $S^{-1}M$ has a natural $S^{-1}A$-module structure, and we have a canonical isomorphism

$$S^{-1}A \otimes_A M \cong S^{-1}M.$$

The functor $M \mapsto S^{-1}M$ is *exact*, which shows that $S^{-1}A$ is a *flat* A-module (recall that an A-module F is called **flat** if the functor

$$M \mapsto F \otimes_A M$$

is exact, cf. [Bour], Chap. I).

The prime ideals of $S^{-1}A$ are the ideals $S^{-1}\mathfrak{p}$, where $\mathfrak{p}$ ranges over the set of prime ideals of A which do not intersect S; if $\mathfrak{p}$ is such an ideal, the preimage of $S^{-1}\mathfrak{p}$ under $A \to S^{-1}A$ is $\mathfrak{p}$.

Example (i).   If $\mathfrak{p}$ is a prime ideal of A, take S to be the complement $A - \mathfrak{p}$ of $\mathfrak{p}$. Then one writes $A_\mathfrak{p}$ and $M_\mathfrak{p}$ instead of $S^{-1}A$ and $S^{-1}M$. The ring $A_\mathfrak{p}$ is a *local ring* with maximal ideal $\mathfrak{p}A_\mathfrak{p}$, whose residue field is the field of fractions of $A/\mathfrak{p}$; the prime ideals of $A_\mathfrak{p}$ correspond bijectively to the prime ideals of A contained in $\mathfrak{p}$.

It is easily seen that, if $M \neq 0$, there exists a prime ideal $\mathfrak{p}$ with $M_\mathfrak{p} \neq 0$ (and one may even choose $\mathfrak{p}$ to be maximal). More generally, *if N is a submodule of M, and x is an element of M, one has $x \in N$ if and only if this is so "locally"*, i.e. the image of x in $M_\mathfrak{p}$ belongs to $N_\mathfrak{p}$ for every prime ideal $\mathfrak{p}$ (apply the above to the module $(N + Ax)/N$.

Example (ii).   If x is a non-nilpotent element of A, take S to be the set of powers of x. The ring $S^{-1}A$ is then $\neq 0$, and so has a prime ideal; whence the existence of a prime ideal of A not containing x. In other words:

Proposition 2.   *The intersection of the prime ideals of A is the set of nilpotent elements of A.*

Corollary.   *Let $\mathfrak{a}$ and $\mathfrak{b}$ be two ideals of A. The following properties are equivalent:*
(1) *Every prime ideal containing $\mathfrak{a}$ contains $\mathfrak{b}$ (i.e. $\mathcal{V}(\mathfrak{a}) \subset \mathcal{V}(\mathfrak{b})$ with the notation of §5 below).*
(2) *For every $x \in \mathfrak{b}$ there exists $n \geq 1$ such that $x^n \in \mathfrak{a}$.*
    If $\mathfrak{b}$ is finitely generated, these properties are equivalent to:
(3) *There exists $m \geq 1$ such that $\mathfrak{b}^m \subset \mathfrak{a}$.*

The implications $(2) \Rightarrow (1)$ and $(3) \Rightarrow (2)$ are clear. The implication $(1) \Rightarrow (2)$ follows from proposition 2, applied to $A/\mathfrak{a}$. If $\mathfrak{b}$ is generated by $x_1, \ldots, x_r$ and if $x_i^n \in \mathfrak{a}$ for every i, the ideal $\mathfrak{b}^m$ is generated by the monomials

$$\mathbf{x}^\mathbf{d} = x_1^{d_1} \cdots x_r^{d_r} \qquad \text{with } \sum d_i = m.$$

If $m > (n-1)r$, one of the d_i's is $\geq n$, hence $\mathbf{x}^\mathbf{d}$ belongs to $\mathfrak{a}$, and we have $\mathfrak{b}^m \subset \mathfrak{a}$. Hence $(2) \Rightarrow (3)$.

Remark.   The set of $x \in A$ such that there exists $n(x) \geq 1$ with $x^{n(x)} \in \mathfrak{a}$ is an ideal, called the **radical** of $\mathfrak{a}$, and denoted by $\text{rad}(\mathfrak{a})$. Condition (2) can then be written as $\mathfrak{b} \subset \text{rad}(\mathfrak{a})$.

4. Noetherian rings and modules

An A-module M is called **noetherian** if it satisfies the following equivalent conditions:
  a) every ascending chain of submodules of M stops;
  b) every non-empty family of submodules of M has a maximal element;
  c) every submodule of M is finitely generated.
    If N is a submodule of M, one proves easily that:

$$M \text{ is noetherian} \iff N \text{ and } M/N \text{ are noetherian}.$$

The ring A is called **noetherian** if it is a noetherian module (when viewed as an A-module), i.e. if every ideal of A is finitely generated. If A is noetherian, so are the rings $A[X_1, \ldots, X_n]$ and $A[[X_1, \ldots, X_n]]$ of polynomials and formal power series over A, see e.g. [Bour], Chap. III, §2, no. 10.
    When A is noetherian, and M is an A-module, conditions a), b), c) above are equivalent to:
  d) M is finitely generated.
    (Most of the rings and modules we shall consider later will be noetherian.)

5. Spectrum   ([Bour], Chap. II, §4)

The **spectrum** of A is the set $\mathrm{Spec}(A)$ of prime ideals of A. If $\mathfrak{a}$ is an ideal of A, the set of $\mathfrak{p} \in \mathrm{Spec}(A)$ such that $\mathfrak{a} \subset \mathfrak{p}$ is written as $\mathcal{V}(\mathfrak{a})$. We have

$$\mathcal{V}(\mathfrak{a} \cap \mathfrak{b}) \;=\; \mathcal{V}(\mathfrak{ab}) \;=\; \mathcal{V}(\mathfrak{a}) \cup \mathcal{V}(\mathfrak{b}) \quad \text{and} \quad \mathcal{V}(\Sigma \mathfrak{a}_i) \;=\; \cap \mathcal{V}(\mathfrak{a}_i).$$

The $\mathcal{V}(\mathfrak{a})$ are the closed sets for a topology on $\mathrm{Spec}(A)$, called the **Zariski topology**. If A is noetherian, the space $\mathrm{Spec}(A)$ is noetherian: every increasing sequence of open subsets stops.
    If F is a closed set $\neq \emptyset$ of $\mathrm{Spec}(A)$, the following properties are equivalent:
  i) F is **irreducible**, i.e. it is not the union of two closed subsets distinct from F;
  ii) there exists $\mathfrak{p} \in \mathrm{Spec}(A)$ such that $F = \mathcal{V}(\mathfrak{p})$, or, equivalently, such that F is the closure of $\{\mathfrak{p}\}$.

    Now let M be a finitely generated A-module, and $\mathfrak{a} = \mathrm{Ann}(M)$ its **annihilator**, i.e. the set of $a \in A$ such that $a_M = 0$, where a_M denotes the endomorphism of M defined by a.

Proposition 3. *If $\mathfrak{p}$ is a prime ideal of A, the following properties are equivalent:*
a) $M_\mathfrak{p} \neq 0$;
b) $\mathfrak{p} \in \mathcal{V}(\mathfrak{a})$.

Indeed, the hypothesis that M is finitely generated implies that the annihilator of the $A_\mathfrak{p}$-module $M_\mathfrak{p}$ is $\mathfrak{a}_\mathfrak{p}$, whence the result.

The set of $\mathfrak{p} \in \operatorname{Spec}(A)$ having properties a) and b) is denoted by $\operatorname{Supp}(M)$, and is called the **support** of M. It is a closed subset of $\operatorname{Spec}(A)$.

Proposition 4.
a) If $0 \to M' \to M \to M'' \to 0$ *is an exact sequence of finitely generated A-modules, then*
$$\operatorname{Supp}(M) = \operatorname{Supp}(M') \cup \operatorname{Supp}(M'').$$
b) *If P and Q are submodules of a finitely generated module M, then*
$$\operatorname{Supp}(M/(P \cap Q)) = \operatorname{Supp}(M/P) \cup \operatorname{Supp}(M/Q).$$
c) *If M and N are two finitely generated modules, then*
$$\operatorname{Supp}(M \otimes_A N) = \operatorname{Supp}(M) \cap \operatorname{Supp}(N).$$

Assertions a) and b) are clear. Assertion c) follows from cor. 2 to prop. 1, applied to the localizations $M_\mathfrak{p}$ and $N_\mathfrak{p}$ of M and N at $\mathfrak{p}$.

Corollary. *If M is a finitely generated module, and $\mathfrak{r}$ an ideal of A, then*
$$\operatorname{Supp}(M/\mathfrak{r}M) = \operatorname{Supp}(M) \cap \mathcal{V}(\mathfrak{r}).$$

This follows from c) since $M/\mathfrak{r}M = M \otimes_A A/\mathfrak{r}$.

6. The noetherian case

In this section and the following ones, *we suppose that A is noetherian*.

The spectrum $\operatorname{Spec}(A)$ of A is then a quasi-compact noetherian space. If F is a closed subset of $\operatorname{Spec}(A)$, every irreducible subset of F is contained in a maximal irreducible subset of F, and these are closed in F; each such subset is called an **irreducible component** of F. The set of irreducible components of F is *finite*; the union of these components is equal to F.

The irreducible components of $\mathrm{Spec}(A)$ are the $\mathcal{V}(\mathfrak{p})$, where $\mathfrak{p}$ ranges over the (finite) set of *minimal prime ideals* of A. More generally, let M be a finitely generated A-module, with annihilator $\mathfrak{a}$. The irreducible components of $\mathrm{Supp}(M)$ are the $\mathcal{V}(\mathfrak{p})$, where $\mathfrak{p}$ ranges over the set of prime ideals having any of the following equivalent properties:

  i) $\mathfrak{p}$ contains $\mathfrak{a}$, and is minimal with this property;
 ii) $\mathfrak{p}$ is a minimal element of $\mathrm{Supp}(M)$;
iii) the module $M_\mathfrak{p}$ is $\neq 0$, and of finite length over the ring $A_\mathfrak{p}$.

(Recall that a module is of **finite length** if it has a Jordan-Hölder sequence; in the present case, this is equivalent to saying that the module is finitely generated, and that its support contains only maximal ideals.)

7. Associated prime ideals   ([Bour], Chap. IV, §1)

Recall that A is assumed to be noetherian.

Let M be a finitely generated A-module, and $\mathfrak{p} \in \mathrm{Spec}(A)$. A prime ideal $\mathfrak{p}$ of A is said to be **associated to** M if M contains a submodule isomorphic to $A/\mathfrak{p}$, equivalently if there exists an element of M whose annihilator is equal to $\mathfrak{p}$. The set of prime ideals associated to M is written as $\mathrm{Ass}(M)$.

Proposition 5.   *Let P be the set of annihilators of the nonzero elements of M. Then every maximal element of P is a prime ideal.*

Let m be an element $\neq 0$ of M whose annihilator $\mathfrak{p}$ is a maximal element of P. If $xy \in \mathfrak{p}$ and $x \notin \mathfrak{p}$, then $xm \neq 0$, the annihilator of xm contains $\mathfrak{p}$, and is therefore equal to $\mathfrak{p}$, since $\mathfrak{p}$ is maximal in P. Since $yxm = 0$, we have $y \in \mathrm{Ann}(xm) = \mathfrak{p}$, which proves that $\mathfrak{p}$ is prime.

Corollary 1.   If $M \neq 0$, then $\mathrm{Ass}(M) \neq \emptyset$.

Indeed, P is then non-empty, and therefore has a maximal element, since A is noetherian.

Corollary 2.   *There exists an increasing sequence $(M_i)_{0 \leq i \leq n}$ of submodules of M, with $M_0 = 0$ and $M_n = M$, such that, for $1 \leq i \leq n$, M_i/M_{i-1} is isomorphic to $A/\mathfrak{p}_i$, with $\mathfrak{p}_i \in \mathrm{Spec}(A)$.*

If $M \neq 0$, corollary 1 shows that there exists a submodule M_1 of M isomorphic to $A/\mathfrak{p}_1$, with $\mathfrak{p}_1$ prime. If $M_1 \neq M$, the same argument, applied to M/M_1, proves the existence of a submodule M_2 of M containing

M_1 and such that M_2/M_1 is isomorphic to $A/\mathfrak{p}_2$, with $\mathfrak{p}_2$ prime, and so on. We obtain an increasing sequence (M_i); in view of the noetherian character of M, this sequence stops; whence the desired result.

Exercise.   Deduce from corollary 1 (or prove directly) that the natural map

$$M \to \prod_{\mathfrak{p} \in \mathrm{Ass}(M)} M_{\mathfrak{p}}$$

is injective.

Proposition 6.   *Let S be a subset of A closed under multiplication and containing 1; let $\mathfrak{p} \in \mathrm{Spec}(A)$ be such that $S \cap \mathfrak{p} = \emptyset$. In order that the prime ideal $S^{-1}\mathfrak{p}$ of $S^{-1}A$ is associated to $S^{-1}M$, it is necessary and sufficient that $\mathfrak{p}$ is associated to M.*

(In other words, Ass is compatible with localization.)

If $\mathfrak{p} \in \mathrm{Ass}(M)$, there is an element $m \in M$ whose annihilator is $\mathfrak{p}$; the annihilator of the element $m/1$ of $S^{-1}M$ is $S^{-1}\mathfrak{p}$; this shows that $S^{-1}\mathfrak{p} \in \mathrm{Ass}(S^{-1}M)$.

Conversely, suppose that $S^{-1}\mathfrak{p}$ is the annihilator of an element m/s of $S^{-1}M$, with $m \in M$, $s \in S$. If $\mathfrak{a}$ is the annihilator if m, then $S^{-1}\mathfrak{a} = S^{-1}\mathfrak{p}$, which implies $\mathfrak{a} \subset \mathfrak{p}$, cf. §3, and also implies the existence of $s' \in S$ with $s'\mathfrak{p} \subset \mathfrak{a}$. One checks that the annihilator of $s'm$ is $\mathfrak{p}$, whence $\mathfrak{p} \in \mathrm{Ass}(M)$.

Theorem 1.   *Let $(M_i)_{0 \leq i \leq n}$ be an increasing sequence of submodules of M, with $M_0 = 0$ and $M_n = M$, such that, for $1 \leq i \leq n$, M_i/M_{i-1} is isomorphic to $A/\mathfrak{p}_i$, with $\mathfrak{p}_i \in \mathrm{Spec}(A)$, cf. corollary 2 to proposition 5. Then*

$$\mathrm{Ass}(M) \subset \{\mathfrak{p}_1, \ldots, \mathfrak{p}_n\} \subset \mathrm{Supp}(M),$$

and these three sets have the same minimal elements.

Let $\mathfrak{p} \in \mathrm{Spec}(A)$. Then $M_{\mathfrak{p}} \neq 0$ if and only if one of $(A/\mathfrak{p}_i)_{\mathfrak{p}}$ is $\neq 0$, i.e. if and only if $\mathfrak{p}$ contains one of $\mathfrak{p}_i$. This shows that $\mathrm{Supp}(M)$ contains $\{\mathfrak{p}_1, \ldots, \mathfrak{p}_n\}$, and that these two sets have the same minimal elements.

On the other hand, if $\mathfrak{p} \in \mathrm{Ass}(M)$, the module M contains a submodule N isomorphic to $A/\mathfrak{p}$. Let i be the smallest index such that $N \cap M_i \neq 0$; if m is a nonzero element of $N \cap M_i$, the module Am is isomorphic to $A/\mathfrak{p}$, and maps injectively into $M_i/M_{i-1} \cong A/\mathfrak{p}_i$; this implies $\mathfrak{p} = \mathfrak{p}_i$, whence the inclusion $\mathrm{Ass}(M) \subset \{\mathfrak{p}_i, \ldots, \mathfrak{p}_n\}$.

Finally, if $\mathfrak{p}$ is a minimal element of $\mathrm{Supp}(M)$, the support $\mathrm{Supp}(M_{\mathfrak{p}})$ of the localization of M at $\mathfrak{p}$ is reduced to the unique maximal ideal $\mathfrak{p}A_{\mathfrak{p}}$ of $A_{\mathfrak{p}}$. As $\mathrm{Ass}(M_{\mathfrak{p}})$ is non-empty (cor. 1 to prop. 5) and contained in $\mathrm{Supp}(M_{\mathfrak{p}})$, we necessarily have $\mathfrak{p}A_{\mathfrak{p}} \in \mathrm{Ass}(M_{\mathfrak{p}})$, and proposition 6 (applied to $S = A - \mathfrak{p}$) shows that $\mathfrak{p} \in \mathrm{Ass}(M)$, which proves the theorem.

Corollary . $\mathrm{Ass}(M)$ *is finite.*

A non-minimal element of $\mathrm{Ass}(M)$ is sometimes called **embedded**.

Proposition 7.   *Let* $\mathfrak{a}$ *be an ideal of* A. *The following properties are equivalent:*

  i) *there exists* $m \in M$, $m \neq 0$, *such that* $\mathfrak{a}m = 0$;

 ii) *for each* $x \in \mathfrak{a}$, *there exists* $m \in M$, $m \neq 0$, *such that* $xm = 0$;

iii) *there exists* $\mathfrak{p} \in \mathrm{Ass}(M)$ *such that* $\mathfrak{a} \subset \mathfrak{p}$;

iv) $\mathfrak{a}$ *is contained in the union of the ideals* $\mathfrak{p} \in \mathrm{Ass}(M)$.

The equivalence of i) and iii) follows from prop. 5 and the noetherian property of A. The equivalence of iii) and iv) follows from the finiteness of $\mathrm{Ass}(M)$, together with the following lemma:

Lemma 1.   *Let* $\mathfrak{a}, \mathfrak{p}_1, \dots, \mathfrak{p}_n$ *be ideals of a commutative ring* R. *If the* $\mathfrak{p}_i$ *are prime, and if* $\mathfrak{a}$ *is contained in the union of the* $\mathfrak{p}_i$, *then* $\mathfrak{a}$ *is contained in one of the* $\mathfrak{p}_i$.

(It is not necessary to suppose that all the $\mathfrak{p}_i$ are prime; it suffices that $n - 2$ among them are so, cf. [Bour], Chap. II, §1, prop. 2.)

We argue by induction on n, the case $n = 1$ being trivial. We can suppose that the $\mathfrak{p}_i$ do not have any relation of inclusion among them (otherwise, we are reduced to the case of $n - 1$ prime ideals). We have to show that, if $\mathfrak{a}$ is not contained in any of the $\mathfrak{p}_i$, there exists $x \in \mathfrak{a}$ which does not belong to any of the $\mathfrak{p}_i$. According to the induction hypothesis, there exists $y \in \mathfrak{a}$ such that $y \notin \mathfrak{p}_i$, $1 \leq i \leq n-1$. If $y \notin \mathfrak{p}_n$, we take $x = y$. If $y \in \mathfrak{p}_n$, we take $x = y + zt_1 \cdots t_{n-1}$, with

$$z \in \mathfrak{a}, \quad z \notin \mathfrak{p}_n, \quad \text{and} \quad t_i \in \mathfrak{p}_i, \quad t_i \notin \mathfrak{p}_n.$$

One checks that x satisfies our requirement.

Let us go back to the proof of prop. 7. The implication i) $\Rightarrow$ ii) is trivial, and ii) $\Rightarrow$ iv) follows from what has already been shown (applied to the case of a principal ideal). The four properties i), ii), iii), iv) are therefore equivalent.

Corollary . *For an element of A to be a zero-divisor, it is necessary and sufficient that it belongs to an ideal $\mathfrak{p}$ in $\mathrm{Ass}(A)$.*

This follows from prop. 7 applied to $M = A$.

Proposition 8. *Let $x \in A$ and let x_M be the endomorphism of M defined by x. The following conditions are equivalent:*
  i) *x_M is nilpotent;*
  ii) *x belongs to the intersection of $\mathfrak{p} \in \mathrm{Ass}(M)$ (or of $\mathfrak{p} \in \mathrm{Supp}(M)$, which amounts to the same according to theorem 1).*

If $\mathfrak{p} \in \mathrm{Ass}(M)$, M contains a submodule isomorphic to $A/\mathfrak{p}$; if x_M is nilpotent, its restriction to this submodule is also nilpotent, which implies $x \in \mathfrak{p}$; whence i) $\Rightarrow$ ii).

Conversely, suppose ii) holds, and let (M_i) be an increasing sequence of submodules of M satisfying the conditions of cor. 2 to prop. 5. According to th. 1, x belongs to every corresponding prime ideal $\mathfrak{p}_i$, and we have $x_M(M_i) \subset M_{i-1}$ for each i, whence ii).

Corollary . *Let $\mathfrak{p} \in \mathrm{Spec}(A)$. Suppose $M \neq 0$. For $\mathrm{Ass}(M) = \{\mathfrak{p}\}$, it is necessary and sufficient that x_M is nilpotent (resp. injective) for every $x \in \mathfrak{p}$ (resp. for every $x \notin \mathfrak{p}$).*

This follows from propositions 7 and 8.

Proposition 9. *If N is a submodule of M, one has:*
$$\mathrm{Ass}(N) \subset \mathrm{Ass}(M) \subset \mathrm{Ass}(N) \cup \mathrm{Ass}(M/N).$$

The inclusion $\mathrm{Ass}(N) \subset \mathrm{Ass}(M)$ is clear. If $\mathfrak{p} \in \mathrm{Ass}(M)$, let E be a submodule of M isomorphic to $A/\mathfrak{p}$. If $E \cap N = 0$, E is isomorphic to a submodule of M/N, and $\mathfrak{p}$ belongs to $\mathrm{Ass}(M/N)$. If $E \cap N \neq 0$, and x is a nonzero element of $E \cap N$, the submodule Ax is isomorphic to $A/\mathfrak{p}$, and $\mathfrak{p}$ belongs to $\mathrm{Ass}(N)$. This shows that $\mathrm{Ass}(M)$ is contained in $\mathrm{Ass}(N) \cup \mathrm{Ass}(M/N)$.

Proposition 10. *There exists an embedding*
$$M \to \prod_{\mathfrak{p} \in \mathrm{Ass}(M)} E(\mathfrak{p}),$$
where, for every $\mathfrak{p} \in \mathrm{Ass}(M)$, $E(\mathfrak{p})$ is such that $\mathrm{Ass}(E(\mathfrak{p})) = \{\mathfrak{p}\}$.

For every $\mathfrak{p} \in \mathrm{Ass}(M)$, choose a submodule $Q(\mathfrak{p})$ of M such that

$\mathfrak{p} \notin \mathrm{Ass}(Q(\mathfrak{p}))$, and maximal for that property. One has $Q(\mathfrak{p}) \neq M$. Define $E(\mathfrak{p})$ as $M/Q(\mathfrak{p})$. If $\mathfrak{q} \in \mathrm{Spec}(A)$ is distinct from $\mathfrak{p}$, $E(\mathfrak{p})$ cannot contain a submodule $M'/Q(\mathfrak{p})$ isomorphic to $A/\mathfrak{q}$, since we would have $\mathfrak{p} \notin \mathrm{Ass}(M')$ by proposition 9, and this would contradict the maximality of $Q(\mathfrak{p})$. Hence $\mathrm{Ass}(E(\mathfrak{p})) \subset \{\mathfrak{p}\}$, and equality holds since $E(\mathfrak{p}) \neq 0$, cf. cor. 1 to prop. 5. The same corollary shows that the intersection of the $Q(\mathfrak{p})$, for $\mathfrak{p} \in \mathrm{Ass}(M)$, is 0, hence the canonical map

$$M \to \prod E(\mathfrak{p})$$

is injective.

Remark. When $\mathfrak{p}$ is a minimal element of $\mathrm{Ass}(M)$, the kernel of the map $M \to E(\mathfrak{p})$ is equal to the kernel of the localizing map $M \to M_{\mathfrak{p}}$, hence is independent of the chosen embedding. There is no such uniqueness for the embedded primes.

8. Primary decompositions ([Bour], Chap. IV, §2)

Let A, M be as above. If $\mathfrak{p} \in \mathrm{Spec}(A)$, a submodule Q of M is called a $\mathfrak{p}$-**primary** submodule of M if $\mathrm{Ass}(M/Q) = \{\mathfrak{p}\}$.

Proposition 11.   *Every submodule N of M can be written as an intersection:*

$$N = \bigcap_{\mathfrak{p} \in \mathrm{Ass}(M/N)} Q(\mathfrak{p})$$

where $Q(\mathfrak{p})$ is a $\mathfrak{p}$-primary submodule of M.

This follows from prop. 10, applied to M/N.

Remark.   Such a decomposition $N = \bigcap Q(\mathfrak{p})$ is called a **reduced** (or minimal) **primary decomposition** of N in M. The elements of $\mathrm{Ass}(M/N)$ are sometimes called the **essential** prime ideals of N in M.

The most important case is the one where $M = A$, $N = \mathfrak{q}$, with $\mathfrak{q}$ being an ideal of A. One then says that $\mathfrak{q}$ is $\mathfrak{p}$-primary if it is $\mathfrak{p}$-primary in A; one then has $\mathfrak{p}^n \subset \mathfrak{q} \subset \mathfrak{p}$ for some $n \geq 1$, and every element of $A/\mathfrak{q}$ which does not belong to $\mathfrak{p}/\mathfrak{q}$ is a non-zero-divisor.

(The reader should be warned that, if $\mathfrak{a}$ is an ideal of A, an element of $\mathrm{Ass}(A/\mathfrak{a})$ is often said to be "associated to $\mathfrak{a}$", cf. e.g. [Eis], p. 89. We shall try not to use this somewhat confusing terminology.)

Chapter II. Tools

A: Filtrations and Gradings

(For more details, the reader is referred to [Bour], Chap. III.)

1. Filtered rings and modules

Definition 1. A **filtered ring** is a ring A given with a family $(A_n)_{n \in \mathbf{Z}}$ of ideals satisfying the following conditions:

$$A_0 = A, \qquad A_{n+1} \subset A_n, \qquad A_p A_q \subset A_{p+q}.$$

A **filtered module** over the filtered ring A is an A-module M given with a family $(M_n)_{n \in \mathbf{Z}}$ of submodules satisfying the following conditions:

$$M_0 = M, \qquad M_{n+1} \subset M_n, \qquad A_p M_q \subset M_{p+q}.$$

[Note that these definitions are more restrictive than those of [Bour], *loc. cit.*]

The filtered modules form an additive category F_A, the morphisms being the A-linear maps $u : M \to N$ such that $u(M_n) \subset N_n$. If P is an A-submodule of the filtered module M, the **induced filtration** on P is the filtration (P_n) defined by the formula $P_n = P \cap M_n$. Similarly, the **quotient filtration** on $N = M/P$ is the filtration (N_n) where the submodule $N_n = (M_n + P)/P$ is the image of M_n.

In F_A, the notions of injective (resp. surjective) morphisms are the usual notions. Every morphism $u : M \to N$ admits a kernel $\mathrm{Ker}(u)$ and a cokernel $\mathrm{Coker}(u)$: the underlying modules of $\mathrm{Ker}(u)$ and $\mathrm{Coker}(u)$ are the usual kernel and cokernel, together with the induced filtration and the quotient filtration. We similarly define $\mathrm{Im}(u) = \mathrm{Ker}(N \to \mathrm{Coker}(u))$ and $\mathrm{Coim}(u) = \mathrm{Coker}(\mathrm{Ker}(u) \to M)$. We have the canonical factorization:

$$\mathrm{Ker}(u) \to M \to \mathrm{Coim}(u) \xrightarrow{\theta} \mathrm{Im}(u) \to N \to \mathrm{Coker}(u),$$

where θ is bijective. One says that u is a **strict morphism** if θ is an isomorphism of filtered modules; it amounts to the same as saying that $u(M_n) = N_n \cap u(M)$ for each $n \in \mathbf{Z}$. There exist bijective morphisms that are not isomorphisms (F_A is not an *abelian* category).

Examples of filtrations.

a) If $\mathfrak{m}$ is an ideal of A, the $\mathfrak{m}$-**adic filtration** of A (resp. of the A-module M) is the filtration for which $A_n = \mathfrak{m}^n$ for $n \geq 1$ (resp. $M_n = \mathfrak{m}^n M$ for $n \geq 1$).

b) Let A be a filtered ring, N a filtered A-module, and M an A-module. The submodules $\mathrm{Hom}_A(M, N_n)$ of $\mathrm{Hom}_A(M, N)$ define on $\mathrm{Hom}_A(M, N)$ a filtered module structure.

2. Topology defined by a filtration

If M is a filtered A-module, the M_n are a basis of neighborhoods of 0 for a topology on M compatible with its group structure (cf. Bourbaki, *TG III*). This holds in particular for A itself, which thus becomes a *topological ring*; similarly, M is a topological A-module.

If $\mathfrak{m}$ is an ideal of A, the $\mathfrak{m}$-**adic topology** on an A-module M is the topology defined by the $\mathfrak{m}$-adic filtration of M.

Proposition 1. *Let N be a submodule of a filtered module M. The closure $\overline{N}$ of N is equal to $\bigcap(N + M_n)$.*

Indeed, saying that x does not belong to $\overline{N}$ means that there exists $n \in \mathbf{Z}$ such that $(x + M_n) \cap N = \emptyset$, i.e. that x does not belong to $N + M_n$.

Corollary.   *M is Hausdorff if and only if $\bigcap M_n = 0$.*

3. Completion of filtered modules

If M is a filtered A-module, we write $\hat{M}$ for its Hausdorff completion; this is an $\hat{A}$-module, isomorphic to $\varprojlim M/M_n$. If we set

$$\hat{M}_n = \text{Ker}(\hat{M} \to M/M_n),$$

$\hat{M}$ becomes a filtered $\hat{A}$-module, and $\hat{M}/\hat{M}_n = M/M_n$; $\hat{M}_n$ is the completion of M_n, with the filtration induced by that of M.

Proposition 2. *Let M be a filtered module, Hausdorff and complete. A series $\sum x_n$, $x_n \in M$, converges in M if and only if its general term x_n tends toward zero.*

The condition is obviously necessary. Conversely, if $x_n \to 0$, there exists for every p an integer $n(p)$ such that $n \geq n(p) \Rightarrow x_n \in M_p$. Then $x_n + x_{n+1} + \ldots + x_{n+k} \in M_p$ for every $k \geq 0$, and the Cauchy criterion applies.

Proposition 3. *Let A be a ring and $\mathfrak{m}$ an ideal of A. If A is Hausdorff and complete for the $\mathfrak{m}$-adic topology, the ring of formal power series $A[[X]]$ is Hausdorff and complete for the $(\mathfrak{m}, X)$-adic topology.*

The ideal $(\mathfrak{m}, X)^n$ consists of the series $a_0 + a_1 X + \ldots + a_k X^k + \ldots$ such that $a_p \in \mathfrak{m}^{n-p}$ for $0 \leq p \leq n$. The topology defined by these ideals in $A[[X]]$ is therefore the topology of pointwise convergence of the coefficients a_i; i.e., $A[[X]]$ is isomorphic (as a topological group) to the product $A^{\mathbf{N}}$, which is indeed Hausdorff and complete.

Proposition 4. *Let $\mathfrak{m}_1, \ldots, \mathfrak{m}_k$ be pairwise distinct maximal ideals of the ring A, and let $\mathfrak{r} = \mathfrak{m}_1 \cap \ldots \cap \mathfrak{m}_k$. Then there is a canonical isomorphism*

$$\hat{A} = \prod_{1 \leq i \leq k} \hat{A}_{\mathfrak{m}_i}$$

where $\hat{A}$ is the completion of A for the $\mathfrak{r}$-adic topology, and where $\hat{A}_{\mathfrak{m}_i}$ is the Hausdorff completion of $A_{\mathfrak{m}_i}$ for the $\mathfrak{m}_i A_{\mathfrak{m}_i}$-adic topology.
[There is an analogous result for modules.]

As the $\mathfrak{m}_i$, $1 \leq i \leq k$, are pairwise distinct, we have

$$A/\mathfrak{r}^n = A/(\mathfrak{m}_1^n \cap \ldots \cap \mathfrak{m}_k^n) = \prod_{1 \leq i \leq k} A_{\mathfrak{m}_i}/\mathfrak{m}_i^n A_{\mathfrak{m}_i}.$$

[We are using here a variant of Bézout's lemma: if $\mathfrak{a}_1,\dots,\mathfrak{a}_k$ are ideals of A such that $\mathfrak{a}_i + \mathfrak{a}_j = A$ for $i \neq j$, the map $A \to \prod A/\mathfrak{a}_i$ is surjective, with kernel equal to $\mathfrak{a}_1 \cdots \mathfrak{a}_k$, cf. e.g. [Bour], Chap. II, §1, no. 2.]
   Hence:

$$\hat{A} = \varprojlim A/\mathfrak{r}^n = \prod_{1 \leq i \leq k} \varprojlim(A_{\mathfrak{m}_i}/\mathfrak{m}_i^n A_{\mathfrak{m}_i}) = \prod \hat{A}_{\mathfrak{m}_i}.$$

Remark.   The proposition applies to the case of a *semi-local* ring A, taking for $\mathfrak{m}_i$ the set of maximal ideals of A; the ideal $\mathfrak{r}$ is then the *radical* of A.

4. Graded rings and modules

Definition 2.   A **graded ring** is a ring A given with a direct sum decomposition

$$A = \bigoplus_{n \in \mathbf{Z}} A_n,$$

where the A_n are additive subgroups of A such that $A_n = \{0\}$ if $n < 0$ and $A_p A_q \subset A_{p+q}$. A **graded module** over the graded ring A is an A-module M given with a direct sum decomposition

$$M = \bigoplus_{n \in \mathbf{Z}} M_n,$$

where the M_n are additive subgroups of M such that $M_n = \{0\}$ if $n < 0$ and $A_p M_q \subset M_{p+q}$.

   Now let M be a filtered module over a filtered ring A. We write $\operatorname{gr}(M)$ for the direct sum $\bigoplus \operatorname{gr}_n(M)$, where $\operatorname{gr}_n(M) = M_n/M_{n+1}$. The canonical maps from $A_p \times M_q$ to M_{p+q} define, by passing to quotients, bilinear maps from $\operatorname{gr}_p(A) \times \operatorname{gr}_q(M)$ to $\operatorname{gr}_{p+q}(M)$, whence a bilinear map from $\operatorname{gr}(A) \times \operatorname{gr}(M)$ to $\operatorname{gr}(M)$.
   In particular, for $M = A$, we obtain a *graded ring* structure on $\operatorname{gr}(A)$; this is the **graded ring associated to the filtered ring** A. Similarly, the map $\operatorname{gr}(A) \times \operatorname{gr}(M) \to \operatorname{gr}(M)$ provides $\operatorname{gr}(M)$ with a $\operatorname{gr}(A)$-graded module structure. If $u : M \to N$ is a morphism of filtered modules, u defines, by passing to quotients, homomorphisms

$$\operatorname{gr}_n(u) : M_n/M_{n+1} \to N_n/N_{n+1},$$

whence a homomorphism $\operatorname{gr}(u) : \operatorname{gr}(M) \to \operatorname{gr}(N)$.

Example.   Let k be a ring, and let $A = k[[X_1,\dots,X_r]]$ be the algebra of formal power series over k in the indeterminates $X_1,\dots,X_r$. Let

$\mathfrak{m} = (X_1, \ldots, X_r)$, and provide A with the $\mathfrak{m}$-adic filtration. The graded ring $\mathrm{gr}(A)$ associated to A is the polynomial algebra $k[X_1, \ldots, X_n]$, graded by total degree.

The modules M, $\hat{M}$ and $\mathrm{gr}(M)$ have similar properties. First:

Proposition 5.   *The canonical maps* $M \to \hat{M}$ *and* $A \to \hat{A}$ *induce isomorphisms* $\mathrm{gr}(M) = \mathrm{gr}(\hat{M})$ *and* $\mathrm{gr}(A) = \mathrm{gr}(\hat{A})$.

This is clear.

Proposition 6.   *Let* $u : M \to N$ *be a morphism of filtered modules. We suppose that* M *is complete,* N *is Hausdorff, and* $\mathrm{gr}(u)$ *is surjective. Then* u *is a surjective strict morphism, and* N *is complete.*

Let n be an integer, and let $y \in N_n$. We construct a sequence $(x_k)_{k \geq 0}$ of elements of M_n such that

$$x_{k+1} \equiv x_k \mod M_{n+k} \quad \text{and} \quad u(x_k) \equiv y \mod N_{n+k}.$$

We proceed by induction starting with $x_0 = 0$. If x_k has been constructed, we have $u(x_k) - y \in N_{n+k}$ and the surjectivity of $\mathrm{gr}(u)$ shows that there exists $t_k \in M_{n+k}$ such that $u(t_k) \equiv u(x_k) - y \mod N_{n+k+1}$; we take $x_{k+1} = x_k - t_k$. Let x be one of the limits in M of the Cauchy sequence (x_k); as M_n is closed, we have $x \in M_n$, and $u(x) = \lim u(x_k)$ is equal to y. Therefore $u(M_n) = N_n$, which shows that u is a surjective strict morphism. The topology of N is a quotient of that of M, and it is therefore a complete module.

Corollary 1.   *Let* A *be a complete filtered ring,* M *a Hausdorff filtered* A-*module,* $(x_i)_{i \in I}$ *a finite family of elements of* M, *and* (n_i) *a finite family of integers such that* $x_i \in M_{n_i}$. *Let* $\overline{x_i}$ *be the image of* x_i *in* $\mathrm{gr}_{n_i}(M)$. *If the* $\overline{x_i}$ *generate the* $\mathrm{gr}(A)$-*module* $\mathrm{gr}(M)$, *then the* x_i *generate* M, *and* M *is complete.*

Let $E = A^I$, and let E_n be the subgroup of E which consists of $(a_i)_{i \in I}$ such that $a_i \in A_{n-n_i}$ for each $i \in I$. This defines a filtration of E, and the associated topology is the product topology of A^I. Let $u : E \to M$ be the homomorphism given by:

$$u((a_i)) = \sum a_i x_i.$$

This is a morphism of filtered modules, and the hypothesis made on the $\overline{x_i}$ amounts to saying that $\mathrm{gr}(u)$ is surjective. Hence the result according to proposition 6. [The proof shows also that $M_n = \sum A_{n-n_i} x_i$ for each integer n .]

Corollary 2.   If M is a Hausdorff filtered module over the complete filtered ring A, and if $\mathrm{gr}(M)$ is a finitely generated (resp. noetherian) $\mathrm{gr}(A)$-module, then M is finitely generated (resp. noetherian, and each of its submodules is closed).

Corollary 1 shows that, if $\mathrm{gr}(M)$ is finitely generated, then M is complete and finitely generated. Moreover, if N is a submodule of M, with the induced filtration, then $\mathrm{gr}(N)$ is a graded $\mathrm{gr}(A)$-submodule of $\mathrm{gr}(M)$; thus if $\mathrm{gr}(M)$ is noetherian, $\mathrm{gr}(N)$ is finitely generated, and N is finitely generated and complete (therefore closed since M is Hausdorff); hence M is noetherian.

Corollary 3.   Let $\mathfrak{m}$ be an ideal of the ring A. Suppose that $A/\mathfrak{m}$ is noetherian, $\mathfrak{m}$ is finitely generated, and A is Hausdorff and complete for the $\mathfrak{m}$-adic topology. Then A is noetherian.

Indeed, if $\mathfrak{m}$ is generated by $x_1, \dots, x_r$, then $\mathrm{gr}(A)$ is a quotient of the polynomial algebra $(A/\mathfrak{m})[X_1, \dots, X_r]$, and therefore is noetherian. The corollary above then shows that A is noetherian.

Proposition 7.   If the filtered ring A is Hausdorff, and if $\mathrm{gr}(A)$ is a domain, then A is a domain.

Indeed, let x and y be two nonzero elements of A. We may find n, m such that $x \in A_n - A_{n+1}$, $y \in A_m - A_{m+1}$; the elements x and y then define nonzero elements of $\mathrm{gr}(A)$; since $\mathrm{gr}(A)$ is a domain, the product of these elements is nonzero, and a fortiori we have $xy \neq 0$, whence the result.

One can similarly show that if A is Hausdorff, noetherian, if every principal ideal of A is closed, and if $\mathrm{gr}(A)$ is a domain and is integrally closed, then A is a domain and is integrally closed (cf. for example [ZS], vol. II, p.250 or [Bour], Chap. V, §1). In particular, if k is a noetherian domain, and is integrally closed, the same is true for $k[X]$ and for $k[[X]]$.

Note also that, if k is a complete nondiscrete valuation field, the local ring $k\langle\langle X_1, \dots, X_r \rangle\rangle$ of convergent series with coefficients in k is noetherian and factorial (that may be seen via Weierstrass "preparation theorem").

5. Where everything becomes noetherian again — q-adic filtrations

From now on, the rings and modules considered are assumed to be *noetherian*. We consider such a ring A and an ideal $\mathfrak{q}$ of A; we provide A with its $\mathfrak{q}$-adic filtration.

Let M be an A-module filtered by (M_n) with $\mathfrak{q}M_n \subset M_{n+1}$ for every $n \geq 0$. We associate to it the **graded group** $\overline{M}$ which is the *direct sum of the* M_n, $n \geq 0$; in particular, $\overline{A} = \bigoplus \mathfrak{q}^n$. The canonical maps $A_p \times M_q \to M_{p+q}$ extend to a bilinear map from $\overline{A} \times \overline{M}$ to $\overline{M}$; this defines a graded A-algebra structure on $\overline{A}$, and a graded $\overline{A}$-module structure on $\overline{M}$ [in algebraic geometry, $\overline{A}$ corresponds to blowing up at the subvariety defined by $\mathfrak{q}$, cf. e.g. [Eis], §5.2].

Since $\mathfrak{q}$ is finitely generated, $\overline{A}$ is an A-algebra generated by a finite number of elements, and thus is in particular a noetherian ring.

Proposition 8.   *The following three properties are equivalent:*
 (a) *We have* $M_{n+1} = \mathfrak{q}M_n$ *for* n *sufficiently large.*
 (b) *There exists an integer* m *such that* $M_{m+k} = \mathfrak{q}^k M_m$ *for* $k \geq 0$.
 (c) $\overline{M}$ *is a finitely generated* $\overline{A}$-module.

The equivalence of (a) and (b) is trivial. If (b) holds for an integer m, it is clear that $\overline{M}$ is generated by $\sum_{i \leq m} M_i$, whence it is finitely generated; hence (c). Conversely, if $\overline{M}$ is generated by homogeneous elements of degree n_i, it is clear that we have $M_{n+1} = \mathfrak{q}M_n$ for $n \geq \sup n_i$; whence (c) $\Rightarrow$ (a).

Definition .   The filtration (M_n) of M is called $\mathfrak{q}$-**good** if it satisfies the equivalent conditions of prop. 8 (i.e., we have $M_{n+1} \supset \mathfrak{q}M_n$ for all n, with equality for almost all n).

Theorem 1 (Artin-Rees).   *If* P *is a submodule of* M, *the filtration induced on* P *by the* $\mathfrak{q}$-adic filtration of M is $\mathfrak{q}$-good. In other words, there exists an integer m such that*

$$P \cap \mathfrak{q}^{m+k} M = \mathfrak{q}^k (P \cap \mathfrak{q}^m M) \qquad \text{for all } k \geq 0.$$

We clearly have $\overline{P} \subset \overline{M}$; since $\overline{M}$ is finitely generated, and $\overline{A}$ is noetherian, $\overline{P}$ is finitely generated, which proves the theorem.

[This presentation of the Artin-Rees Lemma is due to Cartier; it is reproduced in [Bour], Chap. III, §3.]

Corollary 1.   *Every A-linear map $u : M \to N$ is a strict homomorphism of topological groups (in the sense of Bourbaki, TG III) when M and N are given the $\mathfrak{q}$-adic topology.*

It is trivial that the $\mathfrak{q}$-adic topology of $u(M)$ is a quotient of that of M, and theorem 1 implies that it is induced by that of N.

Corollary 2.   *The canonical map $\hat{A} \otimes_A M \to \hat{M}$ is bijective, and the ring $\hat{A}$ is A-flat.*

The first assertion is obvious if M is free. In the general case, choose an exact sequence:
$$L_1 \to L_0 \to M \to 0$$
where the L_i are free. We have a commutative diagram with exact rows:

$$
\begin{array}{ccccccc}
\hat{A} \otimes_A L_1 & \longrightarrow & \hat{A} \otimes_A L_0 & \longrightarrow & \hat{A} \otimes_A M & \longrightarrow & 0 \\
\phi_1 \downarrow & & \phi_0 \downarrow & & \phi \downarrow & & \\
\hat{L}_1 & \longrightarrow & \hat{L}_0 & \longrightarrow & \hat{M} & \longrightarrow & 0.
\end{array}
$$

Since ϕ_0 and ϕ_1 are bijective, so is ϕ. Further, since the functor $M \mapsto \hat{M}$ is left exact, so is the functor $M \mapsto \hat{A} \otimes_A M$ (in the category of finitely generated modules — therefore also in the category of all modules), which means that $\hat{A}$ is A-flat.

Corollary 3.   *If we identify the Hausdorff completion of a submodule N of M with a submodule of $\hat{M}$, we have the formulae:*
$$\hat{N} = \hat{A}N, \quad \hat{N}_1 + \hat{N}_2 = (N_1 + N_2)\widehat{}, \quad \hat{N}_1 \cap \hat{N}_2 = (N_1 \cap N_2)\widehat{}.$$

We leave the proof to the reader; it uses only the noetherian hypothesis and the fact that $\hat{A}$ is flat. In particular, corollary 3 remains valid when we replace the functor $M \mapsto \hat{M}$ by the "localization" functor $M \mapsto S^{-1}M$, where S is a multiplicatively closed subset of A.

Corollary 4.   *The following properties are equivalent:*
 (i) *$\mathfrak{q}$ is contained in the radical $\mathfrak{r}$ of A.*
 (ii) *Every finitely generated A-module is Hausdorff for the $\mathfrak{q}$-adic topology.*
 (iii) *Every submodule of a finitely generated A-module is closed for the $\mathfrak{q}$-adic topology.*

(i) $\Rightarrow$ (ii). Let P be the closure of 0; the $\mathfrak{q}$-adic topology of P is the coarsest topology, whence $P = \mathfrak{q}P$, and since $\mathfrak{q} \subset \mathfrak{r}$, this implies $P = 0$ by Nakayama's lemma.

(ii) $\Rightarrow$ (iii). If N is a submodule of M, the fact that M/N is Hausdorff implies that N is closed.

(iii) $\Rightarrow$ (i). Let $\mathfrak{m}$ be a maximal ideal of A. Since $\mathfrak{m}$ is closed in A, we have $\mathfrak{q} \subset \mathfrak{m}$, whence also $\mathfrak{q} \subset \mathfrak{r}$.

Corollary 5.   *If A is local, and if $\mathfrak{q}$ is distinct from A, we have*

$$\bigcap_{n \geq 0} \mathfrak{q}^n = 0.$$

This follows from corollary 4.

Definition . A **Zariski ring** is a noetherian topological ring whose topology can be defined by the powers of an ideal $\mathfrak{q}$ contained in the radical of the ring. [This condition does not determine $\mathfrak{q}$ in general; but if $\mathfrak{q}'$ satisfies it, we have $\mathfrak{q}^n \subset \mathfrak{q}'$ and $(\mathfrak{q}')^m \subset \mathfrak{q}$ for some suitable integers n and m.]

If A is a Zariski ring, and if M is a finitely generated A-module, the $\mathfrak{q}$-adic topology of M does not depend on the choice of $\mathfrak{q}$ (assuming, of course, that the powers of $\mathfrak{q}$ define the topology of A); it is called the **canonical topology** of M. It is Hausdorff (corollary 4), which allows us to identify M with a submodule of $\hat{M}$. If N is a submodule of M, we have the inclusions $N \subset \hat{N} \subset \hat{M}$ and $N \subset M \subset \hat{M}$, and also $N = \hat{N} \cap M$ (since N is *closed* in M).

B: Hilbert-Samuel Polynomials

1. Review on integer-valued polynomials

The **binomial polynomials** $Q_k(X)$, $k = 0, 1, \ldots$ are:

$$Q_0(X) = 1,$$
$$Q_1(X) = X,$$
$$\ldots$$
$$Q_k(X) = \binom{X}{k} = \frac{X(X-1)\cdots(X-k+1)}{k!},$$
$$\ldots.$$

They make up a basis of $\mathbf{Q}[X]$. Moreover, if Δ denotes the standard **difference operator**:

$$\Delta f(n) = f(n+1) - f(n),$$

one has $\Delta Q_k = Q_{k-1}$ for $k > 0$.

Lemma 1.   *Let f be an element of $\mathbf{Q}[X]$. The following properties are equivalent:*
   a) f is a $\mathbf{Z}$-linear combination of the binomial polynomials Q_k.
   b) One has $f(n) \in \mathbf{Z}$ for all $n \in \mathbf{Z}$.
   c) One has $f(n) \in \mathbf{Z}$ for all $n \in \mathbf{Z}$ large enough.
   d) Δf has property a), and there is at least one integer n such that $f(n)$ belongs to $\mathbf{Z}$.

The implications a) $\Rightarrow$ b) $\Rightarrow$ c) and a) $\Rightarrow$ d) are clear. Conversely, if d) is true, one may write Δf as $\Delta f = \sum e_k Q_k$, with $e_k \in \mathbf{Z}$, hence $f = \sum e_k Q_{k+1} + e_0$, with $e_0 \in \mathbf{Q}$; the fact that f takes at least one integral value on $\mathbf{Z}$ shows that e_0 is an integer. Hence d) $\Leftrightarrow$ a). To prove that c) $\Rightarrow$ a), one uses induction on the degree of f. By applying the induction assumption to Δf, one sees that Δf has property a), hence f has property d), which is equivalent to a), qed.

A polynomial f having properties a), ... , d) above is called an **integer-valued** polynomial.

If f is such a polynomial, we shall write $e_k(f)$ for the coefficient of Q_k in the decomposition of f:

$$f = \sum e_k Q_k.$$

One has $e_k(f) = e_{k-1}(\Delta f)$ if $k > 0$. In particular, if $\deg(f) \le k$, $e_k(f)$ is equal to the constant polynomial $\Delta^k f$, and we have

$$f(X) = e_k(f)\frac{X^k}{k!} + g(X), \qquad \text{with } \deg(g) < k.$$

If $\deg(f) = k$, one has

$$f(n) \sim e_k(f)\frac{n^k}{k!}, \qquad \text{for } n \to \infty;$$

hence:

$$e_k(f) > 0 \iff f(n) > 0 \text{ for all large enough } n.$$

2. Polynomial-like functions

Let f be a function with values in $\mathbf{Z}$ which is defined, either on $\mathbf{Z}$, or on the set of all integers $\geq n_0$, where n_0 is a given integer. We say that f is **polynomial-like** if there exists a polynomial $P_f(X)$ such that $f(n) = P_f(n)$ for all large enough n. It is clear that P_f is uniquely defined by f, and that it is *integer-valued* in the sense defined above.

Lemma 2.   *The following properties are equivalent:*
   (i) f *is polynomial-like;*
   (ii) Δf *is polynomial-like;*
   (iii) *there exists $r \geq 0$ such that $\Delta^r f(n) = 0$ for all large enough n.*

   It is clear that (i) $\Rightarrow$ (ii) $\Rightarrow$ (iii).

   Assume (ii) is true, so that $P_{\Delta f}$ is well-defined. Let R be an integer-valued polynomial with $\Delta R = P_{\Delta f}$ (such a polynomial exists since $P_{\Delta f}$ is integer-valued). The function $g : n \mapsto f(n) - R(n)$ is such that $\Delta g(n) = 0$ for all large n; hence it takes a constant value e_0 on all large n. One has $f(n) = R(n) + e_0$ for all large n; this shows that f is polynomial-like. Hence (ii) $\Leftrightarrow$ (i).

   The implication (iii) $\Rightarrow$ (i) follows from (ii) $\Rightarrow$ (i) applied r times.

Remark.   If f is polynomial-like, with associated polynomial P_f, we shall say that f is of degree k if P_f is of degree k, and we shall write $e_k(f)$ instead of $e_k(P_f)$.

3. The Hilbert polynomial

Recall that a commutative ring A is **artinian** if it satisfies the following equivalent conditions:
   (i) A has finite length (as a module over itself);
   (ii) A is noetherian, and every prime ideal of A is maximal.

   The radical $\mathfrak{r}$ of such a ring is nilpotent, and $A/\mathfrak{r}$ is a product of a finite number of fields.

   In what follows we consider a graded ring $H = \bigoplus H_n$ having the following properties:
   a) H_0 *is artinian;*
   b) *the ring H is generated by H_0 and by a finite number $(x_1, \ldots, x_r)$ of elements of H_1.*

Thus H is the quotient of the polynomial ring $H_0[X_1, \ldots, X_r]$ by a homogeneous ideal. In particular, H is noetherian.

Let $M = \bigoplus M_n$ be a finitely generated graded H-module. Each M_n is a finitely generated H_0-module, hence has finite length. We may thus define a function $n \mapsto \chi(M, n)$ by:

$$\chi(M, n) = \ell_{H_0}(M_n),$$

where ℓ denotes the length. We have $\chi(M, n) = 0$ when n is small enough. The behavior of $\chi(M, n)$ for $n \to +\infty$ is given by:

Theorem 2 (Hilbert). $\chi(M, n)$ *is a polynomial-like function of* n, *of degree* $\leq r - 1$.

We may assume that $H = H_0[X_1, \ldots, X_r]$.

We use induction on r. If $r = 0$, M is a finitely generated module over H_0 and is therefore of finite length. Hence $M_n = 0$ for n large. Assume now that $r > 0$, and that the theorem has been proved for $r - 1$. Let N and R be the kernel and cokernel of the endomorphism ϕ of M defined by X_r. These are graded modules, and we have exact sequences:

$$0 \to N_n \to M_n \xrightarrow{\phi} M_{n+1} \to R_{n+1} \to 0.$$

Hence:

$$\Delta\chi(M, n) = \chi(M, n+1) - \chi(M, n) = \chi(R, n+1) - \chi(N, n).$$

Since $X_r R = 0$ and $X_r N = 0$, R and N may be viewed as graded modules over $H_0[X_1, \ldots, X_{r-1}]$. By the induction assumption, $\chi(R, n)$ and $\chi(N, n)$ are polynomial-like functions of degree $\leq r - 2$. Hence $\Delta\chi(M, n)$ has the same property; by lemma 2, $\chi(M, n)$ is polynomial-like of degree $\leq r - 1$, qed.

Notation.   The polynomial associated to $n \mapsto \chi(M, n)$ is denoted by $Q(M)$, and called the **Hilbert polynomial** of M. Its value at an integer n is written $Q(M, n)$. One has $Q(M) = 0$ if and only if $\ell(M) < \infty$.

Assume $r \geq 1$. Since $\deg Q(M) \leq r - 1$, the polynomial $\Delta^{r-1}Q(M)$ is a constant, equal to $e_{r-1}(Q(M))$ with the notation of §1. One has $\Delta^{r-1}Q(M) \geq 0$ since $Q(M, n) \geq 0$ for n large. Here is an upper bound for $\Delta^{r-1}Q(M)$:

Theorem 2'.   *Assume that M_0 generates M as an H-module. Then:*
   a) $\Delta^{r-1}Q(M) \leq \ell(M_0)$.
   b) The following properties are equivalent:
      b1) $\Delta^{r-1}Q(M) = \ell(M_0)$;

b2) $\chi(M,n) = \ell(M_0)\binom{n+r-1}{r-1}$ for all $n \geq 0$;

b3) the natural map $M_0 \otimes_{H_0} H_0[X_1,\ldots,X_r] \to M$ is an isomorphism.

Here again we may assume that $H = H_0[X_1,\ldots,X_r]$. Put:

$$\tilde{M} = M_0 \otimes_{H_0} H = M_0[X_1,\ldots,X_r].$$

The natural map $\tilde{M} \to M$ is surjective by assumption. If R is its kernel, we have exact sequences

$$0 \to R_n \to \tilde{M}_n \to M_n \to 0 \qquad (n \geq 0).$$

Hence

$$\ell(M_n) + \ell(R_n) = \ell(\tilde{M}_n) = \ell(M_0)\binom{n+r-1}{r-1}.$$

By comparing highest coefficients, we get

(*) $$\Delta^{r-1}Q(M) = \ell(M_0) - \Delta^{r-1}Q(R).$$

This shows that a) is true. It is clear that b2) $\Leftrightarrow$ b3) $\Rightarrow$ b1). It remains to see that b1) $\Rightarrow$ b3). Formula (*) above shows that it is enough to prove:

(**) If R is a nonzero graded submodule of $\tilde{M} = M_0[X_1,\ldots,X_r]$, then $\Delta^{r-1}Q(R) \geq 1$.

To do that, let

$$0 = M^0 \subset M^1 \subset \ldots \subset M^s = M_0$$

be a Jordan-Hölder series of M_0; put $R^i = R \cap \tilde{M}^i = R \cap M^i[X_1,\ldots,X_r]$ for $i = 0,\ldots,s$. Since $R \neq 0$, one can choose i such that $R^i \neq R^{i-1}$. We have

$$Q(R,n) \geq Q(R^i/R^{i-1},n) \qquad \text{for } n \text{ large enough.}$$

Moreover, R^i/R^{i-1} is a nonzero graded submodule of $M^i/M^{i-1} \otimes_{H_0} H$. The H_0-simple module M^i/M^{i-1} is a 1-dimensional vector space over a quotient field k of H_0. Hence R^i/R^{i-1} may be identified with a graded ideal $\mathfrak{a}$ of the polynomial algebra $k[X_1,\ldots,X_r]$. If f is a nonzero homogeneous element of $\mathfrak{a}$, then $\mathfrak{a}$ contains the principal ideal $f \cdot k[X_1,\ldots,X_r]$, and for large enough n, we have

$$Q(R^i/R^{i-1},n+t) = \ell(\mathfrak{a}_{n+t}) \geq \binom{n+r-1}{r-1} \qquad \text{where } t = \deg(f).$$

Hence $\Delta^{r-1}Q(R^i/R^{i-1}) \geq 1$, and a fortiori $\Delta^{r-1}Q(R) \geq 1$, qed.

4. The Samuel polynomial

Let A be a noetherian commutative ring, and let M be a finitely generated A-module, and $\mathfrak{q}$ an ideal of A. We make the following assumption:

(4.1) $$\ell(M/\mathfrak{q}M) < \infty.$$

This is equivalent to:

(4.2)   *All the elements of* $\operatorname{Supp}(M) \cap \mathcal{V}(\mathfrak{q})$ *are maximal ideals.*

(The most important case for what follows is the case where A is local, with maximal ideal $\mathfrak{m}$, and $\mathfrak{q}$ is such that $\mathfrak{m} \supset \mathfrak{q} \supset \mathfrak{m}^s$ for some $s > 0$.)

Let (M_i) be a $\mathfrak{q}$-good filtration of M (cf. part A, §5). We have

$$M = M_0 \supset M_1 \supset \cdots,$$

$$M_i \supset \mathfrak{q}M_{i-1}, \qquad \text{with equality for large } i.$$

Since $\mathcal{V}(\mathfrak{q}^n) = \mathcal{V}(\mathfrak{q})$ for all $n > 0$, the A-modules $M/\mathfrak{q}^n M$ have finite length, and the same is true for M/M_n since $M_n \supset \mathfrak{q}^n M$. Hence the function

$$f_M : n \mapsto \ell(M/M_n)$$

is well-defined.

Theorem 3 (Samuel).   *The function*

$$f_M : n \mapsto \ell(M/M_n)$$

is polynomial-like.

To prove this we may assume that $\operatorname{Ann}(M) = 0$ (if not, replace A by $A/\operatorname{Ann}(M)$, and replace $\mathfrak{q}$ by its image in $A/\operatorname{Ann}(M)$). Then (4.2) shows that the elements of $\mathcal{V}(\mathfrak{q})$ are maximal ideals, i.e. that $A/\mathfrak{q}$ is artinian. Let

$$H = \operatorname{gr}(A) = \bigoplus \mathfrak{q}^n/\mathfrak{q}^{n+1}$$

be the graded ring of A relative to its $\mathfrak{q}$-adic filtration. The direct sum

$$\operatorname{gr}(M) = \bigoplus M_n/M_{n+1}$$

is a graded H-module. If $M_{n+1} = \mathfrak{q}M_n$ for $n \geq n_0$, $\operatorname{gr}(M)$ is generated by

$$M_0/M_1 \oplus \cdots \oplus M_{n_0}/M_{n_0+1};$$

hence it is finitely generated. By theorem 2, applied to H and to $\operatorname{gr}(M)$, the function $n \mapsto \chi(\operatorname{gr}(M), n) = \ell(M_n/M_{n+1})$ is polynomial-like. Moreover, we have $\Delta f_M(n) = \ell(M/M_{n+1}) - \ell(M/M_n) = \ell(M_n/M_{n+1})$. This shows that Δf_M is polynomial-like; by lemma 2, the same is true for f_M, qed.

Remark.   The integer-valued polynomial P_{f_M} associated to f_M will be denoted by $P((M_i))$, and its value at an integer n will be written

$P((M_i), n)$. The proof above shows that:

(4.3) $$\Delta P((M_i)) = Q(\mathrm{gr}(M)),$$

where $Q(\mathrm{gr}(M))$ is the Hilbert polynomial of the graded module $\mathrm{gr}(M)$.

When (M_i) is the $\mathfrak{q}$-adic filtration of M (i.e. $M_i = \mathfrak{q}^i M$ for all $i \geq 0$), we write $P_\mathfrak{q}(M)$ instead of $P((\mathfrak{q}^i M))$. As a matter of fact, there is not much difference between the general case and the $\mathfrak{q}$-adic one, as the following lemma shows:

Lemma 3. *We have*

$$P_\mathfrak{q}(M) = P((M_i)) + R,$$

where R is a polynomial of degree $\leq \deg(P_\mathfrak{q}(M)) - 1$, whose leading term is ≥ 0.

Indeed, let $m \geq 0$ be such that $M_{n+1} = \mathfrak{q} M_n$ for $n \geq m$. We have

$$\mathfrak{q}^{n+m} M \subset M_{n+m} = \mathfrak{q}^n M_m \subset \mathfrak{q}^n M \subset M_n \qquad (n \geq 0),$$

hence:

$$P_\mathfrak{q}(M, n+m) \geq P((M_i), n+m) \geq P_\mathfrak{q}(M, n) \geq P((M_i), n) \quad \text{for } n \text{ large.}$$

This shows that $P_\mathfrak{q}(M, n) - P((M_i), n)$ is ≥ 0 for n large, and that $P_\mathfrak{q}(M)$ and $P((M_i))$ have the same leading term. Hence the lemma.

From now on, we shall be interested mostly in $P_\mathfrak{q}(M)$ and its leading term.

Proposition 9. *Let $\mathfrak{a} = \mathrm{Ann}(M)$, $B = A/\mathfrak{a}$, and denote the B-ideal $(\mathfrak{a} + \mathfrak{q})/\mathfrak{a}$ by $\mathfrak{p}$. Assume that $\mathfrak{p}$ is generated by r elements $x_1, \ldots, x_r$. Then:*

a) $\deg P_\mathfrak{q}(M) \leq r$.

b) $\Delta^r P_\mathfrak{q}(M) \leq \ell(M/\mathfrak{q}M)$.

c) *We have $\Delta^r P_\mathfrak{q}(M) = \ell(M/\mathfrak{q}M)$ if and only if the natural map*

$$\phi : (M/\mathfrak{q}M)[X_1, \ldots, X_r] \rightarrow \mathrm{gr}(M)$$

is an isomorphism.

(The map ϕ is defined via the homomorphism

$$(B/\mathfrak{p})[X_1, \ldots, X_r] \rightarrow \mathrm{gr}(B)$$

given by the x_i.)

We may assume $\mathfrak{a} = 0$, hence $B = A$, $\mathfrak{p} = \mathfrak{q}$, and $\mathrm{gr}(A)$ is a quotient of the polynomial ring $(A/\mathfrak{q})[X_1, \ldots, X_r]$. The case $r = 0$ is trivial. Assume $r \geq 1$. By (4.3), we have

(4.4) $$\Delta^r P_\mathfrak{q}(M) = \Delta^{r-1} Q(\mathrm{gr}(M)).$$

By theorem 2, $Q(\mathrm{gr}(M))$ has degree $\leq r-1$. Hence $P_{\mathfrak{q}}(M)$ has degree $\leq r$, and a) is true. Assertions b) and c) follow from (4.4) and theorem 2'.

The function $M \mapsto P_{\mathfrak{q}}(M)$ is "almost" additive. More precisely, consider an exact sequence

$$0 \to N \to M \to P \to 0.$$

Since $M/\mathfrak{q}M$ has finite length, the same is true for $N/\mathfrak{q}N$ and $P/\mathfrak{q}P$; hence the polynomials $P_{\mathfrak{q}}(N)$ and $P_{\mathfrak{q}}(P)$ are well-defined.

Proposition 10. *We have*

$$P_{\mathfrak{q}}(M) = P_{\mathfrak{q}}(N) + P_{\mathfrak{q}}(P) - R,$$

where R is a polynomial of degree $\leq \deg P_{\mathfrak{q}}(N) - 1$, whose leading term is ≥ 0.

Indeed, put $N_i = \mathfrak{q}^i M \cap N$. By theorem 1 (Artin-Rees), (N_i) is a $\mathfrak{q}$-good filtration of N, and we have

$$\ell(M/\mathfrak{q}^n M) = \ell(N/N_n) + \ell(P/\mathfrak{q}^n P) \qquad \text{for } n \geq 0,$$

hence

$$P_{\mathfrak{q}}(M) = P((N_i)) + P_{\mathfrak{q}}(P).$$

By lemma 3, applied to N, we have

$$P((N_i)) = P_{\mathfrak{q}}(N) + R,$$

with $\deg R \leq \deg P_{\mathfrak{q}}(N) - 1$, and $R(n) \geq 0$ for n large. The proposition follows.

Notation. If d is an integer $\geq \deg P_{\mathfrak{q}}(M)$, we denote by $e_{\mathfrak{q}}(M, d)$ the integer $\Delta^d P_{\mathfrak{q}}(M)$. Hence:

$$e_{\mathfrak{q}}(M, d) = 0 \qquad \text{if } d > \deg P_{\mathfrak{q}}(M),$$
$$e_{\mathfrak{q}}(M, d) \geq 1 \qquad \text{if } d = \deg P_{\mathfrak{q}}(M).$$

Moreover, if $d = \deg P_{\mathfrak{q}}(M)$, we have

$$(4.5) \qquad P_{\mathfrak{q}}(M, n) \sim e_{\mathfrak{q}}(M, d) \frac{n^d}{d!} \qquad \text{for } n \to +\infty.$$

The following additivity property of $e_{\mathfrak{q}}(M, d)$ is a consequence of prop. 10:

Corollary .   *With the notation of prop. 10, one has:*

$$e_q(M,d) = e_q(N,d) + e_q(P,d) \qquad \text{for } d \geq \deg P_q(M).$$

(This will be useful later to define intersection multiplicities, cf. Chap. V, part A.)

Here are a few more elementary properties of $P_q(M)$:

Proposition 11.   *The degree of $P_q(M)$ depends only on M and on $\text{Supp}(M) \cap \mathcal{V}(q)$.*

We may assume that $\text{Ann}(M) = 0$, and thus $\mathcal{V}(q) = \{\mathfrak{m}_1, \ldots, \mathfrak{m}_s\}$, where the $\mathfrak{m}_i$ are maximal ideals of A. Let q' be such that $\mathcal{V}(q') = \mathcal{V}(q)$. We have to show that $\deg P_q(M) = \deg P_{q'}(M)$. Choose an integer $m > 0$ such that $q^m \subset q'$, cf. Chap. I, cor. to prop. 2. We have $q^{mn} \subset q'^n$ for $n \geq 0$, hence

$$P_q(M, mn) \geq P_{q'}(M, n) \qquad \text{for } n \text{ large.}$$

This implies $\deg P_q(M) \geq \deg P_{q'}(M)$. By exchanging the roles of q and q', we get the reverse inequality, hence $\deg P_q(M) = \deg P_{q'}(M)$.

Remark.   A somewhat similar argument shows that

$$\deg P_q(M) = \deg P_q(M') \quad \text{if} \quad \text{Supp}(M) = \text{Supp}(M').$$

Hence $\deg P_q(M)$ depends only on $\text{Supp}(M) \cap \mathcal{V}(q)$.

Proposition 12.   *Suppose that $\mathcal{V}(q) = \{\mathfrak{m}_1, \ldots, \mathfrak{m}_s\}$, where the $\mathfrak{m}_i$ are maximal ideals. Put $A_i = A_{\mathfrak{m}_i}$, $M_i = M_{\mathfrak{m}_i}$ and $q_i = qA_i$. Then:*

$$P_q(M) = \sum_{i=1}^{s} P_{q_i}(M_i).$$

(Hence the study of $P_q(M)$ can be reduced to the case when A is *local*, and q is *primary* with respect to the maximal ideal of A.)

The proposition follows from the Bézout-type isomorphism

$$A/q^n \cong \prod A_i/q_i^n,$$

which shows that $M/q^n M$ is isomorphic to $\prod M_i/q_i^n M$.

Chapter III. Dimension Theory

(For more details, see [Bour], Chap. VIII.)

A: Dimension of Integral Extensions

1. Definitions

Let A be a ring (commutative, with a unit element). A finite increasing sequence

$$\mathfrak{p}_0 \subset \mathfrak{p}_1 \subset \ldots \subset \mathfrak{p}_r \tag{1}$$

of prime ideals of A, such that $\mathfrak{p}_i \neq \mathfrak{p}_{i+1}$ for $0 \leq i \leq r - 1$, is called a **chain of prime ideals** in A. The integer r is called the **length** of the chain; the ideal $\mathfrak{p}_0$ (resp. $\mathfrak{p}_r$) is called its **origin** (resp. its **extremity**); one sometimes says that the chain (1) **joins** $\mathfrak{p}_0$ **to** $\mathfrak{p}_r$.

The chains with origin $\mathfrak{p}_0$ correspond bijectively to the chains of the ring $A/\mathfrak{p}_0$ with origin (0); similarly, those with extremity $\mathfrak{p}_r$ correspond to those of the local ring $A_{\mathfrak{p}_r}$ with extremity the maximal ideal of that ring. One can therefore reduce most questions concerning chains to the special case of *local domains*.

The **dimension** of A, which is written as $\dim A$ or $\dim(A)$, is defined as the supremum (finite or infinite) of the length of the chains of prime ideals in A. An artinian ring is of dimension zero; the ring $\mathbf{Z}$ is of dimension 1. If k is a field, we shall prove later (proposition 13) that the polynomial ring $k[X_1, \ldots, X_n]$ is of dimension n; it is clear anyway that its dimension is $\geq n$, since it contains the chain of length n:

$$0 \subset (X_1) \subset (X_1, X_2) \subset \ldots \subset (X_1, \ldots, X_n).$$

If $\mathfrak{p}$ is a prime ideal of A, the **height** of $\mathfrak{p}$ is defined as the dimension of the local ring $A_{\mathfrak{p}}$; this is the supremum of the length of the chains of prime ideals of A with extremity $\mathfrak{p}$. If $\mathfrak{a}$ is an ideal of A, the **height** of $\mathfrak{a}$ is defined as the infimum of the heights of the prime ideals containing

$\mathfrak{a}$. If we write $\mathcal{V}(\mathfrak{a})$ for the set of these ideals (cf. Chap. I), then:

$$\text{ht}(\mathfrak{a}) = \inf_{\mathfrak{p} \in \mathcal{V}(\mathfrak{a})} \text{ht}(\mathfrak{p}) \qquad (2)$$

[When $\mathcal{V}(\mathfrak{a}) = \emptyset$, i.e. $\mathfrak{a} = A$, this definition should be interpreted as: $\text{ht}(\mathfrak{a}) = \dim A$.]

If $\mathfrak{p}$ is a prime ideal, we obviously have:

$$\text{ht}(\mathfrak{p}) + \dim A/\mathfrak{p} \leq \dim A, \qquad (3)$$

but equality does not necessarily hold, even when A is a local noetherian domain (cf. [Nag3], p. 203, Example 2).

Proposition 1.   *If* $\mathfrak{a} \subset \mathfrak{a}'$, *we have* $\text{ht}(\mathfrak{a}) \leq \text{ht}(\mathfrak{a}')$.

This is clear.

2. Cohen-Seidenberg first theorem

Let B be a commutative A-algebra. Recall that an element x of B is **integral** over A if it satisfies an "equation of integral dependence":

$$x^n + a_1 x^{n-1} + \ldots + a_n = 0, \qquad \text{with } a_i \in A, \qquad (4)$$

for a suitable $n \geq 1$. This is equivalent to saying that the subalgebra $A[x]$ of B generated by x is a finitely generated A-module.

In what follows, we assume that B *contains* A, *and that every element of* B *is integral over* A (in which case one says that B is **integral** over A).

Lemma 1.   *Suppose* B *is a domain. Then*

$$A \text{ is a field} \iff B \text{ is a field}.$$

Suppose A is a field, and let x be a nonzero element of B. Choose an equation (4) of minimal degree. We have $a_n \neq 0$ because of the minimality property. Then x has an inverse in B, namely:

$$-a_n^{-1}(x^{n-1} + a_1 x^{n-2} + \ldots + a_{n-1}).$$

This shows that B is a field.

Conversely, suppose that B is a field, and let a be a nonzero element of A. Let x be its inverse in B, and choose an equation (4) satisfied by x. Dividing by x^{n-1}, we get:

$$x = -(a_1 + a_2 a + \ldots + a_n a^{n-1}),$$

which shows that x belongs to A; hence A is a field.

Now let $\mathfrak{p}$ and $\mathfrak{p}'$ be prime ideals of A and B respectively. One says that $\mathfrak{p}'$ **lies over** $\mathfrak{p}$ if $\mathfrak{p}' \cap A = \mathfrak{p}$.

Proposition 2.
(a) For each prime ideal $\mathfrak{p}$ of A, there exists a prime ideal $\mathfrak{p}'$ of B which lies over $\mathfrak{p}$.
(b) If $\mathfrak{p}' \subset \mathfrak{p}''$ are two prime ideals of B lying over the same prime ideal $\mathfrak{p}$ of A, we have $\mathfrak{p}' = \mathfrak{p}''$.
(c) If $\mathfrak{p}'$ lies over $\mathfrak{p}$, for $\mathfrak{p}$ to be maximal, it is necessary and sufficient that $\mathfrak{p}'$ is maximal.

Assertion (c) follows from lemma 1, applied to $A/\mathfrak{p} \subset B/\mathfrak{p}'$. Assertion (b) follows from (c) applied to $A_\mathfrak{p} \subset B_\mathfrak{p}$ (we write $B_\mathfrak{p}$ for the ring of fractions of B relative to the multiplicative set $A - \mathfrak{p}$). The same argument shows that it suffices to prove (a) when A is local and $\mathfrak{p}$ maximal; in this case, take any maximal ideal of B for $\mathfrak{p}'$, and apply lemma 1.

Corollary .
(i) If $\mathfrak{p}'_0 \subset \dots \subset \mathfrak{p}'_r$ is a chain of prime ideals of B, then the $\mathfrak{p}_i = \mathfrak{p}'_i \cap A$ form a chain of prime ideals of A.
(ii) Conversely, let $\mathfrak{p}_0 \subset \dots \subset \mathfrak{p}_r$ be a chain of prime ideals of A, and let $\mathfrak{p}'_0$ be lying over $\mathfrak{p}_0$. Then there exists a chain $\mathfrak{p}'_0 \subset \dots \subset \mathfrak{p}'_r$ in B, with origin $\mathfrak{p}'_0$, which lies over the given chain (i.e. we have $\mathfrak{p}'_i \cap A = \mathfrak{p}_i$ for all i).

Part (i) follows from (b) of proposition 2. For (ii), we argue by induction on r, the case $r = 0$ being trivial. If $\mathfrak{p}_0 \subset \dots \subset \mathfrak{p}_{r-1}$ is lifted to $\mathfrak{p}'_0 \subset \dots \subset \mathfrak{p}'_{r-1}$, proposition 2 applied to $A/\mathfrak{p}_{r-1} \subset B/\mathfrak{p}'_{r-1}$, shows that there exists $\mathfrak{p}'_r$ containing $\mathfrak{p}'_{r-1}$ and lying over $\mathfrak{p}_r$.

Proposition 3.   We have $\dim A = \dim B$. If $\mathfrak{a}'$ is an ideal of B, and if $\mathfrak{a} = \mathfrak{a}' \cap A$, we have

$$\mathrm{ht}(\mathfrak{a}') \leq \mathrm{ht}(\mathfrak{a}).$$

The equality $\dim A = \dim B$ follows from the above corollary. As for the inequality about the heights, it is clear when $\mathfrak{a}'$ is a prime ideal, and the general case reduces to that.

3. Cohen-Seidenberg second theorem

Recall that a domain A is **integrally closed** if every element of its field of fractions which is integral over A belongs to A.

Proposition 4. *Let A be an integrally closed domain, let K be its field of fractions, let L be a quasi-Galois extension of K (Bourbaki, Algèbre V §9), let B be the integral closure of A in L, let G be the group of automorphisms of L over K, and let $\mathfrak{p}$ be a prime ideal of A. Then G acts transitively on the set of prime ideals of B lying over $\mathfrak{p}$.*

(Recall that the **integral closure** of A in L is the subring of L made up of the elements which are integral over A.)

First suppose that G is finite, and let $\mathfrak{q}$ and $\mathfrak{q}'$ be two prime ideals of B lying over $\mathfrak{p}$. Then the $g\mathfrak{q}$ (where g belongs to G) lie over $\mathfrak{p}$, and it suffices to show that $\mathfrak{q}'$ is contained in one of them (proposition 2), or equivalently, in their union (cf. Chap. I, lemma 1). Thus let $x \in \mathfrak{q}'$. The element $y = \prod g(x)$ is fixed by G; since L/K is quasi-Galois, this shows that there exists a power q of the characteristic exponent of K with $y^q \in K$. We have $y^q \in K \cap B = A$ (since A is integrally closed). Moreover, $y^q \in \mathfrak{q}' \cap A = \mathfrak{p}$, which shows that y^q is contained in $\mathfrak{q}$. Therefore there exists $g \in G$ such that $g(x) \in \mathfrak{q}$, whence $x \in g^{-1}\mathfrak{q}$, qed.

The general case: Let $\mathfrak{q}$ and $\mathfrak{q}'$ lie over $\mathfrak{p}$. For every subfield M of L, containing K, quasi-Galois and finite over K, let $G(M)$ be the subset of G which consists of $g \in G$ which transform $\mathfrak{q} \cap M$ into $\mathfrak{q}' \cap M$. This is clearly a *compact* subset of G, and is non-empty according to what has been shown. As the $G(M)$ form a decreasing filtered family, their intersection is non-empty, qed.

Proposition 5. *Let A be an integrally closed domain. Let B be a domain containing A, and integral over A. Let $\mathfrak{p}_0 \subset \ldots \subset \mathfrak{p}_r$ be a chain of prime ideals of A, and let $\mathfrak{p}'_r$ lie over $\mathfrak{p}_r$. Then there exists a chain $\mathfrak{p}'_0 \subset \ldots \subset \mathfrak{p}'_r$ of B, lying over the given chain, and with extremity $\mathfrak{p}'_r$.*

(In fact, the proposition is true with the hypothesis " B is a domain" replaced by the following: "the nonzero elements of A are non-zero-divisors in B").

The field of fractions of B is algebraic over the field of fractions K of A. Embed it in a quasi-Galois extension L of K, and let C be the integral closure of A in L. Let $\mathfrak{q}'_r$ be a prime ideal of C lying over $\mathfrak{p}'_r$, and let $\mathfrak{q}_0 \subset \ldots \subset \mathfrak{q}_r$ be a chain of prime ideals of C lying over $\mathfrak{p}_0 \subset \ldots \subset \mathfrak{p}_r$. If G denotes the group of K-automorphisms of L, proposition 4 shows that there exists $g \in G$ such that $gq_r = q'_r$; thus if we set $q'_i = gq_i$, and $p'_i = B \cap q'_r$, it is clear that the chain $p'_0 \subset \ldots \subset p'_r$ satisfies our requirements.

Corollary .   *Let A and B be two rings satisfying the hypotheses of the proposition above; let $\mathfrak{b}$ be an ideal of B, and let $\mathfrak{a} = \mathfrak{b} \cap A$. We have*

$$\mathrm{ht}(\mathfrak{a}) = \mathrm{ht}(\mathfrak{b}).$$

When $\mathfrak{b}$ is a prime ideal, this follows from the proposition. In the general case, let $\mathfrak{p}'$ be a prime containing $\mathfrak{b}$, and let $\mathfrak{p} = \mathfrak{p}' \cap A$. According to the above, we have $\mathrm{ht}(\mathfrak{p}') = \mathrm{ht}(\mathfrak{p}) \geq \mathrm{ht}(\mathfrak{a})$. As $\mathrm{ht}(\mathfrak{b}) = \inf \mathrm{ht}(\mathfrak{p}')$, we have $\mathrm{ht}(\mathfrak{b}) \geq \mathrm{ht}(\mathfrak{a})$, and together with prop. 3, we obtain the equality we want.

B: Dimension in Noetherian Rings

1. Dimension of a module

Let M be an A-module, and let $A_M = A/\mathrm{Ann}(M)$ be the ring of scalar multiplications of M. The **dimension of** M, denoted by $\dim M$, is defined as the dimension of the ring A_M. When M is finitely generated, the prime ideals $\mathfrak{p}$ of A containing $\mathrm{Ann}(M)$ are those which belong to the *support* $\mathrm{Supp}(M)$ of M (cf. Chap. I, §5). Hence $\dim M$ is the supremum of the lengths of chains of prime ideals in $\mathrm{Supp}(M)$ [which we also write as $\dim M = \dim \mathrm{Supp}(M)$]; that is to say,

$$\dim M = \sup \dim A/\mathfrak{p} \quad \text{for } \mathfrak{p} \in \mathrm{Supp}(M).$$

In this formula, we can clearly limit ourselves to the *minimal* prime ideals of $\mathrm{Supp}(M)$.

2. The case of noetherian local rings

From now on, we suppose that A is *local noetherian*; we write $\mathfrak{m}(A)$, or just $\mathfrak{m}$, for its radical. An ideal $\mathfrak{q}$ of A is called an **ideal of definition** of A if it is contained in $\mathfrak{m}$, and if it contains a power of $\mathfrak{m}$ (which is equivalent to saying that $A/\mathfrak{q}$ is of finite length).

Let M be a nonzero finitely generated A-module. If $\mathfrak{q}$ is an ideal of definition of A, $M/\mathfrak{q}M$ is of finite length, which allows us to define the Samuel polynomial $P_{\mathfrak{q}}(M, n)$ of M, cf. Chap. II, part B. The degree of this polynomial is independent of the choice of $\mathfrak{q}$ (Chap. II, proposition 11); we denote it by $d(M)$.

Finally, we write $s(M)$ for the infimum of the integers n such that there exist $x_1, \ldots, x_n \in \mathfrak{m}$ with $M/(x_1, \ldots, x_n)M$ of finite length.

Theorem 1. *We have*
$$\dim M = d(M) = s(M).$$

First a lemma:

Lemma 2. *Let $x \in \mathfrak{m}$, and let $_xM$ be the submodule of M which consists of the elements annihilated by x.*

a) *We have $s(M) \leq s(M/xM) + 1$.*

b) *Let $\mathfrak{p}_i$ be the prime ideals of $\mathrm{Supp}(M)$ such that $\dim A/\mathfrak{p}_i = \dim M$. If $x \notin \mathfrak{p}_i$ for all i, we have $\dim M/xM \leq \dim M - 1$.*

c) *If $\mathfrak{q}$ is an ideal of definition of A, the polynomial*
$$P_{\mathfrak{q}}(_xM) - P_{\mathfrak{q}}(M/xM)$$
is of degree $\leq d(M) - 1$.

Assertions a) and b) are trivial. Assertion c) follows from the exact sequences
$$0 \;\rightarrow\; _xM \;\rightarrow\; M \;\rightarrow\; xM \;\rightarrow\; 0$$
$$0 \;\rightarrow\; xM \;\rightarrow\; M \;\rightarrow\; M/xM \;\rightarrow\; 0$$
to which we apply prop. 10 of Chap. II.

We can now prove theorem 1 by arguing "in circle":

i) $\dim M \leq d(M)$.

We use induction on $d(M)$, starting from the case $d(M) = 0$ which is trivial. Thus suppose $d(M) \geq 1$, and let $\mathfrak{p}_0 \in \mathrm{Supp}(M)$ such that $\dim A/\mathfrak{p}_0 = \dim M$; we can suppose $\mathfrak{p}_0$ is minimal in $\mathrm{Supp}(M)$, and M contains a submodule N isomorphic to $A/\mathfrak{p}_0$; since $d(M) \geq d(N)$, we are reduced to proving our assertion for N.

Thus let $\mathfrak{p}_0 \subset \mathfrak{p}_1 \subset \ldots \subset \mathfrak{p}_n$ be a chain of prime ideals in A with origin $\mathfrak{p}_0$. We have to show that $n \leq d(N)$. This is clear if $n = 0$. If not, we can choose $x \in \mathfrak{p}_1 \cap \mathfrak{m}$, with $x \notin \mathfrak{p}_0$. Since the chain $\mathfrak{p}_1 \subset \ldots \subset \mathfrak{p}_n$ belongs to $\mathrm{Supp}(N/xN)$, lemma 2 shows that $\dim N/xN = \dim N - 1$, and that $d(N/xN) \leq d(N) - 1$, whence our assertion follows in virtue of the induction hypothesis applied to N/xN.

ii) $d(M) \leq s(M)$.

Let $\mathfrak{a} = (x_1, \ldots, x_n)$, with $\mathfrak{a} \subset \mathfrak{m}$, and $M/\mathfrak{a}M$ of finite length. The ideal $\mathfrak{q} = \mathfrak{a} + \mathrm{Ann}(M)$ is then an ideal of definition of A, and therefore $P_\mathfrak{a}(M) = P_\mathfrak{q}(M)$. According to prop. 9 of Chap. II, the degree of $P_\mathfrak{a}(M)$ is $\leq n$, whence $d(M) \leq s(M)$.

iii) $s(M) \leq \dim M$.

We use induction on $n = \dim M$ [which is finite, according to i)]. Suppose $n \geq 1$, and let $\mathfrak{p}_i$ be the prime ideals of $\mathrm{Supp}(M)$ such that $\dim A/\mathfrak{p}_i = n$; the ideals are minimal in $\mathrm{Supp}(M)$, whence there are only finitely many of them. They are not maximal when $n \geq 1$. Thus there exists $x \in \mathfrak{m}$, such that $x \notin \mathfrak{p}_i$ for all i. Lemma 2 shows that $s(M) \leq s(M/xM) + 1$, and $\dim M \geq \dim M/xM + 1$. By the induction hypothesis, we have $s(M/xM) \leq \dim M/xM$, whence the result we want, qed.

The theorem above (due to Krull [Kr2] and Samuel [Sa3]) is the main result of dimension theory. It implies:

Corollary 1.   *We have $\dim \hat{M} = \dim M$.*

It is indeed clear that $d(M)$ is not changed by completion.

Corollary 2.   *The dimension of A is finite; it is equal to the minimal number of elements of $\mathfrak{m}$ that generate an ideal of definition.*

This is the equality $\dim M = s(M)$ for $M = A$.

Corollary 3.   *The prime ideals of a noetherian ring satisfy the descending chain condition.*

By localizing, we are reduced to the local case, where our assertion follows from corollary 2.

Corollary 4.   *Let A be a noetherian ring, let $\mathfrak{p}$ be a prime ideal of A, and let n be an integer. The two conditions below are equivalent:*
(i) $\mathrm{ht}(\mathfrak{p}) \leq n$.
(ii) There exists an ideal $\mathfrak{a}$ of A, generated by n elements, such that $\mathfrak{p}$ is a minimal element of $\mathcal{V}(\mathfrak{a})$.

If (ii) holds, the ideal $\mathfrak{a}A_\mathfrak{p}$ is an ideal of definition of $A_\mathfrak{p}$, whence (i). Conversely, if (i) holds, there exists an ideal of definition $\mathfrak{b}$ of $A_\mathfrak{p}$

generated by n elements x_i/s, $s \in A - \mathfrak{p}$. The ideal $\mathfrak{a}$ generated by the x_i then satisfies (ii). [For $n = 1$, this is Krull's "Hauptidealsatz".]

Corollary 5.   *Let A be a noetherian local ring, and let M be a finitely generated A-module. Let $\mathfrak{p}_i$ be the prime ideals in $\mathrm{Supp}(M)$ such that $A/\mathfrak{p}_i = \dim M$. If $x \in \mathfrak{m}(A)$, we have $\dim(M/xM) \geq \dim M - 1$ and equality holds if and only if x does not belong to any of the $\mathfrak{p}_i$.*

This follows from lemma 2, combined with the equalities
$$\dim M = s(M) \quad \text{and} \quad \dim(M/xM) = s(M/xM).$$

3. Systems of parameters

Let A be as above, and let M be a finitely generated nonzero A-module of dimension n. A family $(x_1, \ldots, x_s)$ of elements of $\mathfrak{m}$ is called a **system of parameters** for M if $M/(x_1, \ldots, x_s)M$ is of finite length, and if $s = n$. According to theorem 1, such systems always exist.

Proposition 6.   *Let $x_1, \ldots, x_k$ be elements of $\mathfrak{m}$. Then:*
$$\dim M/(x_1, \ldots, x_k)M + k \geq \dim M.$$
There is equality if and only if $x_1, \ldots, x_k$ form part of a system of parameters of M.

The inequality follows from lemma 2, applied k times. If equality holds, and if $x_{k+1}, \ldots, x_n$ ($n = \dim M$) is a system of parameters of $M/(x_1, \ldots, x_k)M$, the quotient $M/(x_1, \ldots, x_n)M$ is of finite length, which shows that $x_1, \ldots, x_n$ is a system of parameters of M. Conversely, if $x_1, \ldots, x_n$ is a system of parameters of M, we have $n - k \geq \dim M/(x_1, \ldots, x_k)M$, qed.

Proposition 7.   *Let $x_1, \ldots, x_k \in \mathfrak{m}$. The following conditions are equivalent:*
(a) The x_i form a system of parameters for M.
(b) The x_i form a system of parameters for $\hat{M}$.
(c) The x_i form a system of parameters for $A_M = A/\mathrm{Ann}(M)$.

This is obvious.

C: Normal Rings

1. Characterization of normal rings

A ring A is called **normal** if it is a noetherian domain, and is integrally closed. For example, any principal ideal domain is normal; if K is a field, the ring of formal power series $K[[X_1, \ldots, X_n]]$ and the ring of polynomials $K[X_1, \ldots, X_n]$ are normal rings.

Recall that A is a **discrete valuation ring** if it is a principal ideal domain and has one and only one irreducible element, up to multiplication by an invertible element; it is a normal local ring.

Proposition 8. *Let A be a local noetherian domain, with maximal ideal $\mathfrak{m}$. The following conditions are equivalent:*

(i) A is a discrete valuation ring.

(ii) A is normal and of dimension 1.

(iii) A is normal, and there exists an element $a \neq 0$ of $\mathfrak{m}$ such that $\mathfrak{m} \in \text{Ass}(A/aA)$.

(iv) The ideal $\mathfrak{m}$ is principal and nonzero.

(i) $\Rightarrow$ (ii) is trivial.

(ii) $\Rightarrow$ (iii) because if a is a nonzero element of $\mathfrak{m}$, the ideal $\mathfrak{m}$ is the only prime ideal of A containing aA, and aA is therefore $\mathfrak{m}$-primary.

(iii) $\Rightarrow$ (iv). Since $\mathfrak{m}$ belongs to $\text{Ass}(A/aA)$, there exists $x \in A$, $x \notin aA$, such that $\mathfrak{m}x \subset aA$. Thus $\mathfrak{m}xa^{-1} \subset A$ and $xa^{-1} \notin A$. If the ideal $\mathfrak{m}xa^{-1}$ is contained in $\mathfrak{m}$, then since $\mathfrak{m}$ is finitely generated, we conclude that xa^{-1} is integral over A, which is contrary to the hypothesis of normality. Thus there exists $t \in \mathfrak{m}$ such that $u = txa^{-1}$ is an invertible element of A. If y is an element of $\mathfrak{m}$, we have $y = (yxa^{-1})u^{-1}t$, which shows that $\mathfrak{m} = tA$, whence (iv).

(iv) $\Rightarrow$ (i). If $\mathfrak{m} = tA$, we have $\mathfrak{m}^n = t^n A$, and since $\bigcap \mathfrak{m}^n = 0$, for any nonzero element y of A there exists n such that $y \in \mathfrak{m}^n$ and $y \notin \mathfrak{m}^{n+1}$. Thus $y = t^n u$, with u invertible in A, whence $yA = t^n A$; since any ideal of A is a sum of principal ideals, we conclude that any ideal of A is of the form $t^n A$, whence (i).

Proposition 9 (Krull). *Let A be a noetherian domain. For A to be normal, it is necessary and sufficient that it satisfies the two conditions below:*

(a) If $\mathfrak{p}$ is a prime ideal of height 1 of A, the local ring $A_{\mathfrak{p}}$ is a discrete valuation ring.

(b) For every $a \in A$, $a \neq 0$, and every $\mathfrak{p} \in \mathrm{Ass}(A/aA)$, the height of $\mathfrak{p}$ is equal to 1.

If A is normal, the same is true for $A_\mathfrak{p}$; if $\mathfrak{p}$ is of height 1, we have $\dim(A_\mathfrak{p}) = 1$, and proposition 8 shows that $A_\mathfrak{p}$ is a discrete valuation ring. Moreover, if $a \neq 0$, and if $\mathfrak{p} \in \mathrm{Ass}(A/aA)$, proposition 8, applied to $A_\mathfrak{p}$, shows that $A_\mathfrak{p}$ is a discrete valuation ring, and it is in particular a ring of dimension 1, whence $\mathrm{ht}(\mathfrak{p}) = 1$.

Conversely, suppose (a) and (b) hold, and let K be the field of fractions of A. Let $x = b/a$ be an element of K; suppose that x belongs to each of the $A_\mathfrak{p}$, for $\mathrm{ht}(\mathfrak{p}) = 1$; thus we have $b \in aA_\mathfrak{p}$ for $\mathrm{ht}(\mathfrak{p}) = 1$, and according to (b) this implies $b \in aA_\mathfrak{p}$ for every $\mathfrak{p} \in \mathrm{Supp}(A/aA)$, hence $b \in aA$. This shows that $A = \bigcap A_\mathfrak{p}$ where $\mathfrak{p}$ runs through the prime ideals of A of height 1. Since the $A_\mathfrak{p}$ are normal, so is A, qed.

Remark. The proof shows that condition (b) is equivalent to the formula $A = \bigcap A_\mathfrak{p}$ where $\mathfrak{p}$ runs through the prime ideals of A of height 1.

Corollary . *If A is normal, and if $\mathfrak{p}$ is a prime ideal of height 1 of A, the only primary ideals for $\mathfrak{p}$ are the "symbolic powers" $\mathfrak{p}^{(n)}$, defined by the formula $\mathfrak{p}^{(n)} = \mathfrak{p}^n A_\mathfrak{p} \cap A$, $(n \geq 1)$.*

Indeed, the $\mathfrak{p}$-primary ideals of A correspond bijectively to the $\mathfrak{p}A_\mathfrak{p}$-primary ideals of $A_\mathfrak{p}$, which are obviously of the form $\mathfrak{p}^n A_\mathfrak{p}$, $n \geq 1$, since $A_\mathfrak{p}$ is a discrete valuation ring.

2. Properties of normal rings

For a systematic exposition (in the slightly more general framework of "Krull rings"), we refer to [Kr1], or to [Bour], Chap. VII. We summarize the main results.

Let A be a normal ring, and let K be its field of fractions. If $\mathfrak{p}$ is a prime ideal of A of height 1, the normalized discrete valuation associated to the ring $A_\mathfrak{p}$ is written as $v_\mathfrak{p}$; the elements $x \in A$ such that $v_\mathfrak{p}(x) \geq n$ form the ideal $\mathfrak{p}^{(n)}$. If $x \neq 0$, the ideal Ax is only contained in a finite number of prime ideals of height 1; thus $v_\mathfrak{p}(x) = 0$ for almost all $\mathfrak{p}$, and this relation extends to the elements x of K^*. The valuations $v_\mathfrak{p}$ furthermore satisfy the approximation theorem below:

Proposition 10.   Let $\mathfrak{p}_i$, $1 \le i \le k$, be pairwise distinct prime ideals of height 1 of A, and let $n_i \in \mathbf{Z}$, $1 \le i \le k$. Then there exists $x \in K^*$ such that:

$$v_{\mathfrak{p}_i}(x) = n_i \quad (1 \le i \le k) \quad \text{and} \quad v_{\mathfrak{p}}(x) \ge 0 \quad \text{for } \mathfrak{p} \ne \mathfrak{p}_1, \dots, \mathfrak{p}_k.$$

First suppose each $n_i \ge 0$, and let $S = \bigcap(A - \mathfrak{p}_i)$. Let $B = A_S$. It is clear that the ring B is a semilocal ring, whose maximal ideals are the $\mathfrak{p}_i B$, and the corresponding localizations are the $A_{\mathfrak{p}_i}$. Since these localizations are principal domains, it follows that B is also a principal domain; if x/s with $s \in S$ is a generator of the ideal $\mathfrak{p}_1^{n_1} \cdots \mathfrak{p}_k^{n_k} B$, we see that x satisfies our requirements.

In the general case, we first choose $y \in K^*$ such that the integers $m_i = v_{\mathfrak{p}_i}(y)$ are $\le n_i$. Let $\mathfrak{q}_1, \dots, \mathfrak{q}_r$ be the prime ideals of height 1 of A, other than the $\mathfrak{p}_i$, such that $v_{\mathfrak{q}_j}(y)$ are < 0, and set $s_j = -v_{\mathfrak{q}_j}(y)$. According to the previous part of the proof, there exists $z \in A$ such that $v_{\mathfrak{p}_i}(z) = n_i - m_i$ and $v_{\mathfrak{q}_j}(z) = s_j$. The element $x = yz$ then satisfies our requirements, qed.

An ideal $\mathfrak{a}$ of A is called **divisorial** if its essential prime ideals are all of height 1; by the corollary to prop. 9, it amounts to the same as saying that $\mathfrak{a}$ is of the form $\bigcap \mathfrak{p}_i^{(n_i)}$, with $n_i \ge 0$ and $\mathrm{ht}(\mathfrak{p}_i) = 1$; thus we have $x \in \mathfrak{a}$ if and only if $x \in A$ and $v_{\mathfrak{p}_i}(x) \ge n_i$ for all i. We extend this definition to the nonzero fractional ideals of K with respect to A. The divisorial ideals correspond bijectively to the **divisors** of A, i.e. to the elements of the free abelian group generated by the prime ideals of height 1. Every principal ideal is divisorial, and the corresponding divisor is called **principal**.

A ring is called a **Dedekind ring** if it is normal of dimension ≤ 1. Its prime ideals of height 1 are then maximal; for such a prime $\mathfrak{p}$, we have $\mathfrak{p}^{(n)} = \mathfrak{p}^n$. Every nonzero ideal is divisorial.

A noetherian ring A is called **factorial** if it is normal and its divisorial ideals are principal (moreover it suffices that the prime ideals of height 1 are so); it amounts to the same as saying that any two elements of A have a greatest common divisor. Every nonzero element of A can be decomposed in the standard way:

$$x = u\pi_1^{n_1} \cdots \pi_k^{n_k},$$

where u is invertible and the π_i are irreducible elements; this decomposition is essentially unique.

3. Integral closure

Proposition 11.   *Let A be a normal ring, with field of fractions K, and let L be a finite separable extension of K. The integral closure B of A in L is a normal ring, which is a finitely generated A-module.*

Let $\mathrm{Tr}(y)$ be the trace in the extension L/K of an element y of L. One has $\mathrm{Tr}(y) \in A$ if $y \in B$ since A is normal; moreover, since L/K is separable, the K-bilinear form $\mathrm{Tr}(xy)$ is non-degenerate. Let B^* be the set of $y \in L$ such that $\mathrm{Tr}(xy) \in A$ for all $x \in B$; since B contains a free submodule E of rank $[L:K]$, B^* is contained in E^* which is free, and since $B \subset B^*$, B is a finitely generated A-module; in particular, B is a noetherian ring, whence a normal ring, qed.

Remarks.
1)   The set B^* is a fractional ideal of L with respect to B, which is called the **codifferent** of B **with respect to** A. It is easy to see that it is a *divisorial ideal* of L, and thus it can be determined by localizing at the prime ideals of height 1. We are thus reduced to the case of discrete valuation rings, where one can moreover define discriminant, ramification groups, etc., see e.g. [Se3].
2)   When the extension L/K is no longer assumed to be separable, it may happen that the ring B is not noetherian (and *a fortiori* not a finitely generated A-module); one can find an example in [Nag1].

D: Polynomial Rings

1. Dimension of the ring $A[X_1, \dots, X_n]$

Lemma 3.   *Let A be a ring, let $B = A[X]$, let $\mathfrak{p}' \subset \mathfrak{p}''$ be two distinct prime ideals of B, such that $\mathfrak{p}' \cap A$ and $\mathfrak{p}'' \cap A$ are equal to the same prime ideal $\mathfrak{p}$ of A. Then $\mathfrak{p}' = \mathfrak{p}B$.*

Dividing by $\mathfrak{p}B$, we are reduced to the case where $\mathfrak{p} = 0$. After localizing with respect to $S = A - \{0\}$, we are reduced to the case where A is a field, and the lemma is then obvious, since $A[X]$ is a principal ideal domain.

Proposition 12.   *If* $B = A[X]$, *we have*

$$\dim(A) + 1 \ \leq \ \dim(B) \ \leq \ 2\dim(A) + 1.$$

If $\mathfrak{p}_0 \subset \ldots \subset \mathfrak{p}_r$ is a chain of prime ideals of A, we set $\mathfrak{p}'_i = \mathfrak{p}_i B$, $\mathfrak{p}'_{r+1} = \mathfrak{p}_r B + XB$, and we obtain a chain of prime ideals of B of length $r + 1$. Whence $\dim(B) \geq \dim(A) + 1$.

Now if $\mathfrak{p}'_0 \subset \ldots \subset \mathfrak{p}'_s$ is a chain of prime ideals of B, and if we set $\mathfrak{p}_i = \mathfrak{p}'_i \cap A$, the above lemma shows that we cannot have $\mathfrak{p}_i = \mathfrak{p}_{i+1} = \mathfrak{p}_{i+2}$. We can therefore extract, from the sequence of $\mathfrak{p}_i$, an increasing chain comprising at least $(s+1)/2$ elements, which is to say its length is at least $(s-1)/2$; whence $(s-1)/2 \leq \dim(A)$, i.e. $s \leq 2\dim(A) + 1$, which shows that $\dim(B) \leq 2\dim(A) + 1$.

Remark.   There are examples showing that $\dim(B)$ can effectively take any intermediate value between $\dim(A) + 1$ and $2\dim(A) + 1$; see [J]. Nevertheless, in the noetherian case, we will show that

$$\dim(B) \ = \ \dim(A) + 1,$$

cf. prop. 13.

In the two lemmas below, we set $B = A[X]$.

Lemma 4.   *Let* $\mathfrak{a}$ *be an ideal of* A, *and let* $\mathfrak{p}$ *be a prime ideal of* A *minimal in* $\mathcal{V}(\mathfrak{a})$. *Then* $\mathfrak{p}B$ *is a prime ideal in* B *minimal in* $\mathcal{V}(\mathfrak{a}B)$.

We can clearly suppose $\mathfrak{a} = 0$. If $\mathfrak{p}B$ is not minimal, it strictly contains a prime ideal $\mathfrak{q}$. Since $\mathfrak{p}B \cap A = \mathfrak{p}$ is minimal in A, we necessarily have $\mathfrak{q} \cap A = \mathfrak{p}$, and we obtain a contradiction with lemma 3.

Lemma 5.   *Suppose* A *is noetherian. If* $\mathfrak{p}$ *is a prime ideal of* A, *we have* $\mathrm{ht}(\mathfrak{p}) = \mathrm{ht}(\mathfrak{p}B)$.

Let $n = \mathrm{ht}(\mathfrak{p})$. According to cor. 4 to th. 1, there exists an ideal $\mathfrak{a}$ of A, generated by n elements, such that $\mathfrak{p}$ is a minimal element of $\mathcal{V}(\mathfrak{a})$. According to the previous lemma, $\mathfrak{p}B$ is a minimal element of $\mathcal{V}(\mathfrak{a}B)$, and cor. 4 to th. 1 shows that $\mathrm{ht}(\mathfrak{p}B) \leq n$. The opposite inequality follows from the fact that any chain $\{\mathfrak{p}_i\}$ of prime ideals with extremity $\mathfrak{p}$ defines in B a chain $\{\mathfrak{p}_i B\}$ of the same length and with extremity $\mathfrak{p}B$.

Proposition 13. *If A is noetherian, we have*

$$\dim(A[X_1,\dots,X_n]) \ = \ \dim(A) + n.$$

It obviously suffices to prove the result for $A[X]$. We already know that $\dim(A[X]) \geq \dim(A) + 1$, and it amounts to prove the reverse inequality. Thus let $\mathfrak{p}_0' \subset \dots \subset \mathfrak{p}_r'$ be a chain of prime ideals of $B = A[X]$, and let $\mathfrak{p}_i = \mathfrak{p}_i' \cap A$. If the $\mathfrak{p}_i$ are distinct, we have $r \leq \dim(A)$. If not, let j be the largest integer such that $\mathfrak{p}_j = \mathfrak{p}_{j+1}$. According to lemma 3, we have $\mathfrak{p}_j' = \mathfrak{p}_j B$, whence (lemma 4) $\mathrm{ht}(\mathfrak{p}_j') = \mathrm{ht}(\mathfrak{p}_j)$, and since $\mathrm{ht}(\mathfrak{p}_j') \geq j$, we have $\mathrm{ht}(\mathfrak{p}_j) \geq j$. But $\mathfrak{p}_j \subset \mathfrak{p}_{j+2} \subset \dots \subset \mathfrak{p}_r$ is a chain of prime ideals in A. Thus $r - (j+1) + \mathrm{ht}(\mathfrak{p}_j) \leq \dim(A)$, whence $r - 1 \leq \dim(A)$, qed.

2. The normalization lemma

In what follows, k denotes a field. A k-algebra A is called **finitely generated** if it is generated (as a k-algebra) by a finite number of elements; i.e. if there exists an integer $n \geq 0$ and a surjective homomorphism

$$k[X_1,\dots,X_n] \to A.$$

Theorem 2 (Normalization lemma). *Let A be a finitely generated k-algebra, and let $\mathfrak{a}_1 \subset \dots \subset \mathfrak{a}_p$ be an increasing sequence of ideals of A, with $\mathfrak{a}_p \neq A$. Then there exist an integer $n \geq 0$ and elements $x_1,\dots,x_n$ of A, algebraically independent over k, such that*
  a) A is integral over $B = k[x_1,\dots,x_n]$;
  b) for each i, $1 \leq i \leq p$, there exists an integer $h(i) \geq 0$ such that $\mathfrak{a}_i \cap B$ is generated by $(x_1,\dots,x_{h(i)})$.

We first observe that it suffices to prove the theorem *when A is a polynomial algebra* $k[Y_1,\dots,Y_m]$. Indeed, we can write A as a quotient of such an algebra A' by an ideal $\mathfrak{a}_0'$; write $\mathfrak{a}_i'$ for the preimage of $\mathfrak{a}_i$ in A', and let x_i' be elements of A' satisfying the conditions of the theorem relative to the sequence $\mathfrak{a}_0' \subset \mathfrak{a}_1' \subset \dots \subset \mathfrak{a}_p'$. Then it is clear that the images of $x_{i-h(0)}'$ in A, where $i > h(0)$, satisfy the desired conditions.

Thus, in all that follows, we suppose that $A = k[Y_1,\dots,Y_m]$, and we argue by induction on p.

A) $p = 1$.
    We distinguish two cases:

A1) The ideal $\mathfrak{a}_1$ is a principal ideal, generated by $x_1 \notin k$.

We have $x_1 = P(Y_1, \ldots, Y_m)$, where P is a non constant polynomial. Let us show that, for a suitable choice of integers $r_i > 0$, the ring A is integral over $B = k[x_1, x_2, \ldots, x_m]$, with

$$x_i = Y_i - Y_1^{r_i} \qquad (2 \le i \le m).$$

For this, it suffices to show that Y_1 is integral over B. But Y_1 satisfies the equation

$$P(Y_1, x_2 + Y_1^{r_2}, \ldots, x_m + Y_1^{r_m}) - x_1 = 0. \qquad (*)$$

If we write P as a sum of monomials

$$P = \sum a_p Y^p, \qquad \text{where } p = (p_1, \ldots, p_m), \text{ and } a_p \ne 0,$$

equation $(*)$ becomes

$$\sum a_p Y_1^{p_1} (x_2 + Y_1^{r_2})^{p_2} \cdots (x_m + Y_1^{r_m})^{p_m} - x_1 = 0.$$

Set $f(p) = p_1 + r_2 p_2 + \ldots + r_m p_m$, and suppose the r_i are chosen such that the $f(p)$ are all distinct (for example, it suffices to take $r_i = s^i$, with $s > \sup(p_j)$). Then there is one and only one system $p = (p_1, \ldots, p_m)$ such that $f(p)$ is maximum, and the equation is written as:

$$a_p Y_1^{f(p)} + \sum_{j < f(p)} Q_j(x) Y_1^j = 0,$$

which shows that Y_1 is indeed integral over B.

Hence $k(Y_1, \ldots, Y_m)$ is algebraic over $k(x_1, \ldots, x_m)$, which implies that the x_i are algebraically independent, and B is isomorphic to $k[X_1, \ldots, X_m]$. Moreover, $\mathfrak{a}_1 \cap B = (x_1)$; indeed, every element $q \in \mathfrak{a}_1 \cap B$ can be written as $q = x_1 q'$, with $q' \in A \cap k(x_1, \ldots, x_m)$, and we have $A \cap k(x_1, \ldots, x_m) = k[x_1, \ldots, x_m]$ since this ring is integrally closed; whence $q' \in B$, which completes the proof of properties a) and b) in this case.

A2) The general case.

We argue by induction on m, the case $m = 0$ (and also $m = 1$) being trivial. Clearly we can suppose $\mathfrak{a}_1 \ne 0$. Thus let x_1 be a nonzero element of $\mathfrak{a}_1$; this is not a constant since $\mathfrak{a}_1 \ne A$. According to what has been shown, there exist $t_2, \ldots, t_m$, such that $x_1, t_2, \ldots, t_m$ are algebraically independent over k, that A is integral over $C = k[x_1, t_2, \ldots, t_m]$, and that $x_1 A \cap C = x_1 C$. According to the induction hypothesis, there exist elements $x_2, \ldots, x_m$ of $k[t_2, \ldots, t_m]$ satisfying the conclusions of the theorem for the algebra $k[t_2, \ldots, t_m]$ and for the ideal $\mathfrak{a}_1 \cap k[t_2, \ldots, t_m]$. One checks that $x_1, x_2, \ldots, x_m$ satisfy our requirements.

B) Passing from $p - 1$ to p.

Let $t_1, \ldots, t_m$ be elements of A satisfying the conditions of the theorem for the sequence $\mathfrak{a}_1 \subset \ldots \subset \mathfrak{a}_{p-1}$, and let $r = h(p - 1)$. According to A2), there exist elements $x_{r+1}, \ldots, x_m$ of $k[t_{r+1}, \ldots, t_m]$ satisfying the conclusions of the theorem for $k[t_{r+1}, \ldots, t_m]$ and for the ideal

$a_p \cap k[t_{r+1}, \ldots, t_m]$. Setting $x_i = t_i$ for $i \leq r$, we obtain the family we want, qed.

3. Applications. I. Dimension in polynomial algebras

Notation.   If A is a domain which is an algebra over a field k, the transcendence degree over k of the field of fractions of A is written as $\mathrm{alg\,dim}_k A$.

Proposition 14.   *Let A be a domain which is a finitely generated algebra over a field k. We have*

$$\dim(A) = \mathrm{alg\,dim}_k A.$$

According to th. 2, there exists a subalgebra B of A which is isomorphic to a polynomial algebra $k[X_1, \ldots, X_n]$ such that A is integral over B. According to prop. 3, we have $\dim(A) = \dim(B)$, and according to prop. 13, we have $\dim(B) = n$; moreover, if we let L and K denote the fields of fractions of A and B, we have

$$\mathrm{alg\,dim}_k L = \mathrm{alg\,dim}_k K = n,$$

since L is algebraic over K. Whence the proposition.

Variant.   Instead of applying prop. 13, we can apply th. 2 to a chain of prime ideals of A. We deduce that the length of this chain is less than or equal to n (with $B = k[X_1, \ldots, X_n]$) and we conclude as above.

Corollary 1.   *Let A be a finitely generated algebra over a field k, and let $\mathfrak{p}$ be a prime ideal of A. We have*

$$\dim(A/\mathfrak{p}) = \mathrm{alg\,dim}_k(A/\mathfrak{p}).$$

This is obvious.

Corollary 2 ("Nullstellensatz").   *Let A be a finitely generated algebra over a field k, and let $\mathfrak{m}$ be a maximal ideal of A. The field $A/\mathfrak{m}$ is a finite extension of k.*

Since $\mathfrak{m}$ is maximal, we have $\mathrm{coht}(\mathfrak{m}) = 0$, and we apply corollary 1; this shows that $A/\mathfrak{m}$ is algebraic over k; since it is finitely generated, it is a finite extension of k.

Proposition 15.   *Let A be a domain which is a finitely generated algebra over a field k and let $n = \dim(A)$. For every prime ideal $\mathfrak{p}$ of A, we have:*

$$\operatorname{ht}(\mathfrak{p}) + \dim(A/\mathfrak{p}) = n, \qquad i.e. \quad \dim(A_{\mathfrak{p}}) + \dim(A/\mathfrak{p}) = \dim(A).$$

According to theorem 2, there exists a subalgebra B of A, isomorphic to $k[X_1, \ldots, X_n]$, such that A is integral over B, and

$$\mathfrak{p} \cap B = (X_1, \ldots, X_h).$$

Set $\mathfrak{p}' = \mathfrak{p} \cap B$. Since $\mathfrak{p}'$ contains the chain

$$0 \subset (X_1) \subset \ldots \subset (X_1, \ldots, X_h),$$

we have $\operatorname{ht}(\mathfrak{p}') \geq h$, and the opposite inequality follows from the fact that $\mathfrak{p}'$ is generated by h elements, whence $\operatorname{ht}(\mathfrak{p}') = h$. Moreover, $B/\mathfrak{p}' = k[X_{h+1}, \ldots, X_n]$, which shows that $\dim(B/\mathfrak{p}') = n - h$. Since A is integral over B, and B is integrally closed, the Cohen-Seidenberg theorems show that $\operatorname{ht}(\mathfrak{p}) = \operatorname{ht}(\mathfrak{p}')$ and $\dim(A/\mathfrak{p}) = \dim(B/\mathfrak{p}')$. Hence the proposition.

Corollary 1.   *The hypotheses and notation being those of theorem 2, we have*

$$\operatorname{ht}(\mathfrak{a}_i) = h(i) \qquad for \ i = 1, \ldots, p.$$

This is in fact a corollary of the proof.

A chain of prime ideals is called **saturated** if it is not contained in any other chain with the same origin and extremity (in other words, if one cannot interpolate any prime ideal between the elements of the chain); it is called **maximal** if it is not contained in any other chain, or, what amounts to the same, if it is saturated, its origin is a minimal prime ideal, and its extremity is a maximal ideal.

Corollary 2.   *Let A be a domain which is a finitely generated algebra over a field. All the maximal chains of prime ideals of A have the same length, which is $\dim(A)$.*

Let $\mathfrak{p}_0 \subset \mathfrak{p}_1 \subset \ldots \subset \mathfrak{p}_h$ be a maximal chain of prime ideals. Since it is maximal, we have $\mathfrak{p}_0 = 0$, and $\mathfrak{p}_h$ is a maximal ideal of A. We therefore have $\dim(A/\mathfrak{p}_0) = \dim(A)$ and $\dim(A/\mathfrak{p}_h) = 0$. Moreover, since the chain is saturated, one cannot interpolate any prime ideal between $\mathfrak{p}_{i-1}$ and $\mathfrak{p}_i$; thus $\dim(A/\mathfrak{p}_{i-1})_{\mathfrak{p}_i} = 1$, and by prop. 15 we have:

$$\dim(A/\mathfrak{p}_{i-1}) - \dim(A/\mathfrak{p}_i) = 1.$$

As $\dim(A/\mathfrak{p}_0) = \dim(A)$ and $\dim(A/\mathfrak{p}_h) = 0$, we deduce $h = \dim(A)$, qed.

Remarks.
1) Corollary 2 can be split in two parts:
   a) For every maximal ideal $\mathfrak{m}$ of A, we have $\dim(A_\mathfrak{m}) = \dim(A)$.
   b) All the maximal chains of prime ideals of $A_\mathfrak{m}$ have the same length.
   We will see in the following chapter that property b) is true, more generally, for every local ring which is *a quotient of a Cohen-Macaulay ring* (and in particular of a regular local ring).
2) Corollary 2 can also be deduced directly from th. 2.

4. Applications. II. Integral closure of a finitely generated algebra

Proposition 16.   *Let A be a domain which is a finitely generated algebra over a field k, let K be its field of fractions, and let L be a finite extension of K. Then the integral closure B of A in L is a finitely generated A-module (in particular it is a finitely generated k-algebra).*

[Compare this result with prop. 11: we no longer suppose that A is a normal ring, nor that L/K is separable.]

According to th. 2, A is integral over a subalgebra C isomorphic to $k[X_1, \ldots, X_n]$, and B is obviously the integral closure of C in L. It thus suffices to do the proof *when $A = k[X_1, \ldots, X_n]$*. Moreover, being free to extend L, we can suppose that the extension L/K is quasi-Galois; if we write M for the largest purely inseparable extension of K contained in L, the extension L/M is separable. Let D be the integral closure of A in M; if we know that D is finitely generated as a module over A, proposition 11, applied to L/M, shows that B is finitely generated as a module over D, whence over A. Finally, we can thus suppose that the extension L/K is purely inseparable. The extension L is generated by a finite number of elements y_i, and there exists a power q of the characteristic exponent of k such that

$$y_i^q \in K = k(X_1, \ldots, X_n).$$

Let $c_1, \ldots, c_m$ be the coefficients of each of the y_i^q, expressed as rational functions in the X_j. The extension L/K is then contained in L'/K, with:

$$L' = k'(X_1^{q^{-1}}, \ldots, X_n^{q^{-1}}), \quad k' = k(c_1^{q^{-1}}, \ldots, c_m^{q^{-1}}).$$

The integral closure of $A = k[X_1, \ldots, X_n]$ in L' is clearly equal to

$$B' = k'[X_1^{q^{-1}}, \ldots, X_n^{q^{-1}}],$$

and B' is a free A-module with finite basis. Hence B is finitely generated as a module over A, qed.

Remark.   In the terminology of [EGA], Chap. 0, 23.1.1, proposition 16 means that every field is "universally Japanese". According to Nagata, every Dedekind ring of characteristic zero (in particular $\mathbf{Z}$), and every local noetherian complete ring, is universally Japanese (cf. [EGA], Chap. IV, 7.7.4, and [Bour], Chap. IX, §2).

5. Applications. III. Dimension of an intersection in affine space

We want to show that, if V and W are two irreducible subvarieties of an affine space, and if T is an irreducible component of $V \cap W$, we have the inequality:

$$\operatorname{codim}(T) \leq \operatorname{codim}(V) + \operatorname{codim}(W).$$

In algebraic language, this is expressed as:

Proposition 17.   *If $\mathfrak{p}'$ and $\mathfrak{p}''$ are two prime ideals of the polynomial algebra $A = k[X_1, \ldots, X_n]$, where k is a field, and if $\mathfrak{p}$ is a minimal element of $\mathcal{V}(\mathfrak{p}' + \mathfrak{p}'')$, we have*

$$\operatorname{ht}(\mathfrak{p}) \leq \operatorname{ht}(\mathfrak{p}') + \operatorname{ht}(\mathfrak{p}'').$$

We first prove two lemmas:

Lemma 6.   *Let A' and A'' be domains which are finitely generated algebras over k. For every minimal prime ideal $\mathfrak{p}$ of $A' \otimes_k A''$, we have:*

$$\dim(A' \otimes_k A''/\mathfrak{p}) = \dim(A' \otimes_k A'') = \dim(A') + \dim(A'').$$

(In geometric language: the product of two k-irreducible varieties of dimensions r and s decomposes into irreducible varieties of dimension $r + s$.)

Let B' and B'' be k-polynomial algebras of which A' and A'' are integral extensions; let K', K'', L', L'' be the field of fractions of A',

A'', B', B''. We have the diagram of injections:

$$0 \longrightarrow L' \otimes_k L'' \longrightarrow K' \otimes_k K''$$

$$\uparrow \qquad\qquad \uparrow$$

$$0 \longrightarrow B' \otimes_k B'' \longrightarrow A' \otimes_k A''$$

$$\uparrow \qquad\qquad \uparrow$$

$$0 \qquad\qquad 0$$

As K' is L'-free and K'' is L''-free, $K' \otimes_k K''$ is $L' \otimes_k L''$-free; in particular, it is a torsion-free module over the polynomial algebra $B' \otimes_k B''$. The intersection of the prime ideal $\mathfrak{p}$ with $B' \otimes_k B''$ is 0, and the Cohen-Seidenberg theorems show that

$$\begin{aligned} \dim(A' \otimes_k A''/\mathfrak{p}) &= \dim(B' \otimes_k B'') \\ &= \dim(B') + \dim(B'') \\ &= \dim(A') + \dim(A''), \end{aligned}$$

qed.

Lemma 7. *Let A be a k-algebra, let $C = A \otimes_k A$, and let $\phi : C \to A$ be the homomorphism defined by $\phi(a \otimes b) = ab$.*

(i) The kernel $\mathfrak{d}$ of ϕ is the ideal of C generated by the elements

$$1 \otimes a - a \otimes 1, \qquad \text{for } a \in A.$$

(ii) If $\mathfrak{p}'$ and $\mathfrak{p}''$ are two ideals of A, the image by ϕ of the ideal

$$\mathfrak{p}' \otimes A + A \otimes \mathfrak{p}''$$

is equal to $\mathfrak{p}' + \mathfrak{p}''$.

It is clear that $1 \otimes a - a \otimes 1$ belongs to $\mathfrak{d}$ for every $a \in A$. Conversely, if $\sum a_i b_i = 0$, we can write:

$$\sum a_i \otimes b_i = \sum (a_i \otimes 1 - 1 \otimes a_i)(1 \otimes b_i),$$

which shows that $\sum a_i \otimes b_i$ belongs to the ideal generated by the elements $(a_i \otimes 1 - 1 \otimes a_i)$. Assertion (ii) is trivial.

We now proceed to prove the proposition. Set

$$C = A \otimes_k A, \quad D = A/\mathfrak{p}' \otimes_k A/\mathfrak{p}'', \quad \mathfrak{r} = \mathfrak{p}' \otimes A + A \otimes \mathfrak{p}''.$$

We have the exact sequence:

$$0 \to \mathfrak{r} \to C \to D \to 0.$$

Let $\mathfrak{P} = \phi^{-1}(\mathfrak{p})$; it is obviously a minimal prime ideal of $\mathcal{V}(\mathfrak{d} + \mathfrak{r})$, and its image $\mathfrak{Q}$ in D is thus a minimal prime ideal of $\mathcal{V}(\mathfrak{d}')$, where we write $\mathfrak{d}'$ for the image of $\mathfrak{d}$ in D. But lemma 7 shows that $\mathfrak{d}$ is generated by the n elements $X_i \otimes 1 - 1 \otimes X_i$; thus we see that $\mathrm{ht}(\mathfrak{Q}) \leq n$. Let $\mathfrak{Q}_0$ be a minimal prime ideal of D contained in $\mathfrak{Q}$; we have *a fortiori* $\mathrm{ht}(\mathfrak{Q}/\mathfrak{Q}_0) \leq n$. But according to lemma 6, we have

$$\dim(D/\mathfrak{Q}_0) = \dim(A/\mathfrak{p}') + \dim(A/\mathfrak{p}'');$$

since

$$\mathrm{ht}(\mathfrak{Q}/\mathfrak{Q}_0) = \dim(D/\mathfrak{Q}_0) - \dim(D/\mathfrak{Q}),$$

we find

$$n \geq \mathrm{ht}(\mathfrak{Q}/\mathfrak{Q}_0) = \dim(A/\mathfrak{p}') + \dim(A/\mathfrak{p}'') - \dim(A/\mathfrak{p}),$$

hence

$$n - \dim A/\mathfrak{p} \leq n - \dim A/\mathfrak{p}' + n - \dim A/\mathfrak{p}'',$$

i.e.

$$\mathrm{ht}(\mathfrak{p}) \leq \mathrm{ht}(\mathfrak{p}') + \mathrm{ht}(\mathfrak{p}''),$$

qed.

Remark.   The proof above consists in replacing the triple $(A; \mathfrak{p}', \mathfrak{p}'')$ by the triple $(A \otimes_k A; \mathfrak{d}, \mathfrak{r})$. This is called **reduction to the diagonal** (it is the algebraic analogue of the set-theoretic formula $V \cap W = (V \times W) \cap \Delta$). We will see in Chap. V that this method applies to much more general cases, and will allow us to extend the preceding proposition to *all regular rings*.

Chapter IV. Homological Dimension and Depth

A: The Koszul Complex

1. The simple case

Let A be a commutative ring (which is not assumed to be noetherian for the time being) and let x be an element of A. We denote by $K(x)$, or sometimes $K^A(x)$, the following complex:

$$K_i(x) = 0 \quad \text{if } i \neq 0, 1;$$
$$K_1(x) = A;$$
$$K_0(x) = A;$$

the map $d : K_1(x) \to K_0(x)$ is the multiplication by x.

In what follows, we identify $K_0(x)$ with A, and we choose a basis e_x of the A-module $K_1(x)$ such that $d(e_x) = x$. The derivation

$$d : K_1(x) \to K_0(x)$$

is thus defined by the formula:

$$d(ae_x) = ax, \quad \text{if } a \in A.$$

If M is an A-module, we write $K(x, M)$ for the tensor product complex $K(x) \otimes_A M$. Then

$$K(x, M)_n = 0 \quad \text{if } n \neq 0, 1,$$
$$K(x, M)_0 = K_0(x) \otimes_A M \cong M,$$
$$K(x, M)_1 = K_1 \otimes_A M \cong M,$$

and the derivation

$$d : K(x, M)_1 \to K(x, M)_0$$

is defined by the formula

$$d(e_x \otimes m) = xm \quad \text{where } m \in M.$$

The *homology modules* of $K(x, M)$ are:
$$H_0(K(x), M) = M/xM,$$
$$H_1(K(x), M) = \mathrm{Ann}_M(x) = \mathrm{Ker}(x_M : M \to M)$$
$$H_i(K(x), M) = 0 \quad \text{if } i \neq 0, 1.$$
We denote them by $H_i(x, M)$.

Now let L be a complex of A-modules. The homology modules of the complex $K(x) \otimes_A L$ are related to the homology modules of L in the following way:

Proposition 1.   *For every $p \geq 0$, we have an exact sequence:*
$$0 \to H_0(x, H_p(L)) \to H_p(K(x) \otimes_A L) \to H_1(x, H_{p-1}(L)) \to 0.$$

The natural injection $A \to K(x)$ gives an embedding of complexes
$$L = A \otimes_A L \to K(x) \otimes_A L.$$
Similarly, the natural projection $K(x) \to K_1(x) = A$ gives a homomorphism of complexes
$$K(x) \otimes_A L \to L[-1];$$
here, $L[-1]$ is the complex deduced from L by a **shift** of -1 (i.e. $L[-1]_n = L_{n-1}$), together with a sign change on the boundary map.

We thus get an exact sequence of complexes
$$0 \to L \to K(x) \otimes_A L \to L[-1] \to 0,$$
hence a homology exact sequence:
$$\ldots \xrightarrow{d} H_p(L) \to H_p(K(x) \otimes_A L) \to H_p(L[-1]) \xrightarrow{d} H_{p-1}(L) \ldots .$$
The boundary operator d maps $H_p(L[-1]) = H_{p-1}(L)$ into $H_{p-1}(L)$; a simple computation shows that it is equal to scalar multiplication by x. Hence the above exact sequence splits into short exact sequences:
$$0 \to X_p \to H_p(K(x) \otimes_A L) \to Y_{p-1} \to 0,$$
with
$$X_p = \mathrm{Coker}(x : H_p(L) \to H_p(L)) = H_0(x, H_p(L)),$$
$$Y_p = \mathrm{Ker}(x : H_p(L) \to H_p(L)) = H_1(x, H_p(L)),$$
qed.

A complex $L = (L_n)$, $n \geq 0$, is called an **acyclic complex** on M if $H_p(L) = 0$ for $p > 0$ and $H_0(L) = M$. We thus have an exact sequence:
$$\ldots \to L_n \to L_{n-1} \to \ldots \to L_1 \to L_0 \to M \to 0.$$

Corollary . *If L is an acyclic complex on M, and if x is not a zero-divisor in M (i.e. $\mathrm{Ker}(x_M) = 0$), then $K(x) \otimes_A L$ is an acyclic complex on M/xM.*

Indeed, the proposition shows that one has $H_p(K(x) \otimes_A L) = 0$ for $p > 1$, that $H_1(K(x) \otimes_A L) = H_1(K(x) \otimes_A M) = \mathrm{Ker}(x_M) = 0$, and that $H_0(K(x) \otimes_A M)$ is equal to M/xM.

2. Acyclicity and functorial properties of the Koszul complex

If $\mathbf{x} = (x_1, \dots, x_r)$ is a family of elements of A, we let $K(\mathbf{x})$, or $K(x_1, \dots, x_r)$, denote the tensor product complex

$$K(\mathbf{x}) = K(x_1) \otimes_A K(x_2) \otimes_A \cdots \otimes_A K(x_r).$$

Then $K_p(x_1, \dots, x_r)$ is the free A-module generated by the elements $e_{i_1} \otimes \cdots \otimes e_{i_p}$, $i_1 < i_2 < \dots < i_p$, where $e_i = e_{x_i}$, and in particular it is isomorphic to $\bigwedge^p(A^r)$, the p-th exterior product of A^r.

If M is an A-module, we write $K(x_1, \dots, x_r; M)$, or $K(\mathbf{x}, M)$, for the tensor product complex

$$K(x_1, \dots, x_r) \otimes_A M = K(\mathbf{x}) \otimes_A M.$$

The module $K_p(\mathbf{x}, M)$ is thus a direct sum of modules

$$e_{i_1} \otimes_A \cdots \otimes_A e_{i_p} \otimes_A M, \quad \text{where } i_1 < i_2 < \dots < i_p,$$

and the derivation

$$d : K_p(\mathbf{x}, M) \to K_{p-1}(\mathbf{x}, M)$$

is given by the formula: [*]

$$d(e_{i_1} \otimes \cdots \otimes e_{i_p} \otimes m) = \sum_{k=1}^{p} (-1)^{k+1} e_{i_1} \otimes \cdots \otimes \widehat{e_{i_k}} \otimes \cdots \otimes e_{i_p} \otimes (x_{i_k} m).$$

In what follows, we denote by $H_p(\mathbf{x}, M)$ the p-th homology module of the Koszul complex $K(\mathbf{x}, M)$. We have:

$$H_0(\mathbf{x}, M) = M/(x_1, \dots, x_r)M = M/\mathbf{x}M,$$
$$H_r(\mathbf{x}, M) = \{m \in M : x_i m = 0 \text{ for all } i\}.$$

Remark. When one wants to mention the ground ring A, one writes $K^A(x)$, $K^A(\mathbf{x}, M)$, $H_p^A(\mathbf{x}, M)$, etc. Note however that $K(\mathbf{x}, M)$ and

---

[*] We are using here the topologists' convention that a symbol below " $\frown$ " is to be omitted. For instance, for $p = 2$, one has

$$d(e_{i_1} \otimes e_{i_2} \otimes m) = -e_{i_1} \otimes (x_{i_2} m) + e_{i_2} \otimes (x_{i_1} m).$$

$H_p(\mathbf{x}, M)$ depend only on the abelian group M and on the endomorphisms $(x_i)_M$; they do not depend on A.

The two propositions below are concerned with the case where these homology modules are zero for $p > 0$.

Proposition 2.   *Under the preceding hypotheses, if for all i, $1 \le i \le r$, x_i is not a zero-divisor in $M/(x_1, \ldots, x_{i-1})M$, then $H_p(\mathbf{x}, M) = 0$ for $p > 0$.*

The proposition is true if $r = 1$: saying that $H_1(x_1, M) = \operatorname{Ann}_M(x_1)$ is zero is equivalent to saying that x_1 is not a zero-divisor in M.

Thus suppose that $r > 1$ and that the proposition has been proved for the complex $K(x_1, \ldots, x_{r-1}; M)$, and let us prove it for $K(x_1, \ldots, x_r; M)$. The canonical map from $K_0(x_1, \ldots, x_{r-1}; M)$ into $H_0(x_1, \ldots, x_{r-1}; M) = M/(x_1, \ldots, x_{r-1})M$ defines $K(x_1, \ldots, x_{r-1}; M)$ as a complex over $M/(x_1, \ldots, x_{r-1})M$, and the corollary to proposition 1 can be applied.

Proposition 3.   *If, in addition to the preceding hypotheses, we suppose that A is noetherian, that M is finitely generated, and that the x_i belong to the radical $\mathfrak{r}(A)$ of A, then the following properties are equivalent:*
a) $H_p(\mathbf{x}, M) = 0$ for $p \ge 1$.
b) $H_1(\mathbf{x}, M) = 0$.
c) For every i, $1 \le i \le r$, x_i is not a zero-divisor in $M/(x_1, \ldots, x_{i-1})M$.

It remains to show that b) $\Rightarrow$ c), which has already been done if $r = 1$. Assume the result for $K(\mathbf{x}', M)$ where $\mathbf{x}' = (x_1, \ldots, x_{r-1})$ and let us prove it for $K(\mathbf{x}, M)$. By the corollary to proposition 1, we have an exact sequence

$$0 \to H_0(x_r, H_1(\mathbf{x}', M)) \to H_1(\mathbf{x}, M) \to H_1(x_r, H_0(\mathbf{x}, M)) \to 0.$$

Hence $H_1(\mathbf{x}, M) = 0 \Rightarrow H_1(\mathbf{x}', M)/x_r H_1(\mathbf{x}', M) = 0$; by Nakayama's lemma, this shows that $H_1(\mathbf{x}', M) = 0$ and by the induction hypothesis we have property c) for $1 \le i < r$. Moreover, the same exact sequence shows that $H_1(x_r, H_0(\mathbf{x}', M)) = 0$, which is property c) for $i = r$.

Corollary 1.   *Condition c) does not depend on the order of the sequence $\mathbf{x} = \{x_1, \ldots, x_r\}$.*

Remark.   The correspondence between M and $K(\mathbf{x}, M)$ is obviously functorial for a given $\mathbf{x}$, and the functor $M \mapsto K(\mathbf{x}, M)$ is exact. If $0 \to M' \to M \to M'' \to 0$ is an exact sequence, we obtain an exact sequence of complexes:

$$0 \to K(\mathbf{x}, M') \to K(\mathbf{x}, M) \to K(\mathbf{x}, M'') \to 0$$

and an exact sequence of homology:

$$
\begin{aligned}
0 \quad &\to \quad H_r(\mathbf{x}, M') \quad \to \quad H_r(\mathbf{x}, M) \quad \to \quad H_r(\mathbf{x}, M'') \\
&\to \quad H_{r-1}(\mathbf{x}, M') \quad \to \quad \ldots \quad \to \quad H_1(\mathbf{x}, M'') \\
&\to \quad H_0(\mathbf{x}, M') \quad \to \quad H_0(\mathbf{x}, M) \quad \to \quad H_0(\mathbf{x}, M'') \quad \to 0.
\end{aligned}
$$

Moreover, $H_0(\mathbf{x}, M)$ is naturally isomorphic to $A/\mathbf{x} \otimes_A M$ (where $\mathbf{x}$ denotes, as usual, the ideal generated by $x_1, \ldots, x_r$). This isomorphism of functors extends in a unique way to a natural transformation

$$\psi : H_i(\mathbf{x}, M) \to \operatorname{Tor}_i^A(A/\mathbf{x}, M),$$

cf. [CaE], Chap. III.

Corollary 2. *Suppose that conditions* a), b), c) *of proposition 3 are satisfied for* A *(i.e., for* $1 \le i \le r$, x_i *is not a zero-divisor in* $A/(x_1, \ldots, x_{i-1})A$). *Then the map*

$$\psi : H_i(\mathbf{x}, M) \to \operatorname{Tor}_i^A(A/\mathbf{x}, M)$$

is an isomorphism for every i *and for every* A *-module* M (not necessarily finitely generated).

This follows from proposition 3 (applied to the module A), since it implies that $K(\mathbf{x})$ is an A-free resolution of $A/\mathbf{x}$.

Similarly, we have natural maps

$$\phi : \operatorname{Ext}_A^i(A/\mathbf{x}, M) \to H^i(\operatorname{Hom}_A(K(\mathbf{x}), M)).$$

Since $\operatorname{Hom}_A(K(\mathbf{x}), M)$ is isomorphic (with a dimension shift) to $K(\mathbf{x}, M)$ (autoduality of the exterior algebra!), we have

$$H^i(\operatorname{Hom}_A(K(\mathbf{x}), M)) \cong H_{r-i}(\mathbf{x}, M).$$

Hence ϕ may be viewed as a homomorphism of $\operatorname{Ext}_A^i(A/\mathbf{x}, M)$ into $H_{r-i}(\mathbf{x}, M)$; for $i = 0$, this gives the natural isomorphism between $\operatorname{Hom}_A(A/\mathbf{x}, M)$ and $H_r(\mathbf{x}, M)$.

Corollary 3. *Under the same hypothesis as in cor. 2, the map*

$$\phi : \operatorname{Ext}_A^i(A/\mathbf{x}, M) \to H_{r-i}(\mathbf{x}, M)$$

is an isomorphism for every i *and for every* M.

The proof is the same.

Now let B be the polynomial ring in r indeterminates $X_1, \ldots, X_r$, with coefficients in A, i.e. $B = A[X_1, \ldots, X_r]$. Define B-module structures on A and M by the equalities: $X_i a = 0$ if $a \in A$ and $X_i m = x_i m$ if $m \in M$. Then $K^B(X_1, \ldots, X_r)$ gives a free resolution of A and

$$K^A(\mathbf{x}, M) = K^B(X_1, \ldots, X_r; M);$$

thus we have natural isomorphisms:

$$H_i(\mathbf{x}, M) \cong \operatorname{Tor}_i^B(A, M) \cong \operatorname{Ext}_B^{r-i}(A, M).$$

Whence

Proposition 4. *The annihilator of* $H_i(\mathbf{x}, M)$, $-\infty < i < +\infty$, *contains* $\mathbf{x}$ *and* $\operatorname{Ann}(M)$.

Indeed, we have

$$\operatorname{Ann}_B(\operatorname{Tor}_i^B(A, M)) \supset \operatorname{Ann}_B(A) + \operatorname{Ann}_B(M),$$

but $\operatorname{Ann}_B(A) = (X_1, \ldots, X_r)$ and

$$\operatorname{Ann}_B(M) \supset \operatorname{Ann}_A(M) + (X_1 - x_1, \ldots, X_r - x_r).$$

Finally, one shows without difficulty that if S is a multiplicative subset of A, $K(\mathbf{x}, M_S) = K(\mathbf{x}, M)_S$ and $H(\mathbf{x}, M_S) = H(\mathbf{x}, M)_S$. Similarly if A is noetherian and M is finitely generated, and if the A-modules are given the $\mathbf{x}$-adic filtration, then $K(\mathbf{x}, \hat{M}) = \widehat{K(\mathbf{x}, M)}$, $H(\mathbf{x}, \hat{M}) = \widehat{H(\mathbf{x}, M)}$; the relations between $K(\mathbf{x}, M)$ and the Koszul complex of $\operatorname{gr}(M)$ will be studied in the next section.

Remark. For more details on the Koszul complex $K(\mathbf{x}, M)$, see Bourbaki, *Algèbre*, Chap. X, §9, and [Eis], §17.

3. Filtration of a Koszul complex

In this section, A is local noetherian, the ideal $\mathbf{x} = (x_1, \ldots, x_r)$ is contained in the maximal ideal of A, and M is a finitely generated A-module such that $\ell(M/\mathbf{x}M) < \infty$.

The A-modules $H_p(\mathbf{x}, M)$ are finitely generated and they are annihilated by $\mathbf{x} + \operatorname{Ann}(M)$, cf. prop. 4; hence they have finite length, and we may define the **Euler-Poincaré characteristic**

$$\chi(\mathbf{x}, M) = \sum_{p=0}^r (-1)^p \ell(H_p(\mathbf{x}, M)).$$

On the other hand, the Samuel polynomial $P_x(M)$ has degree $\leq r$, and we have

$$P_x(M,n) = e_x(M,r)\frac{n^r}{r!} + Q(n), \quad \text{with } \deg Q < r,$$

where $e_x(M,r) = \Delta^r P_x(M)$, cf. Chap. II, part B, §3.

We want to compare the integers $\chi(\mathbf{x}, M)$ and $e_x(M,r)$:

Theorem 1. $\chi(\mathbf{x}, M) = e_x(M,r)$.

Corollary .   We have $\chi(\mathbf{x}, M) > 0$ if $\dim(M) = r$ and $\chi(\mathbf{x}, M) = 0$ if $\dim(M) < r$.

This follows from the properties of $e_x(M,r)$ proved in Chap. II, part A.

Remark.   Note that $\chi(\mathbf{x}, M)$ is ≥ 0. This is a general property of Koszul complexes in noetherian categories, cf. Appendix II. More generally, we shall see that the "higher Euler-Poincaré characteristics"

$$\chi_i(\mathbf{x}, E) = \sum_{p \geq 0}(-1)^p \ell(H_{i+p}(\mathbf{x}, M))$$

are ≥ 0 for every $i \geq 0$.

Proof of theorem 1.   We do it in several steps:

(3.1)   *Filtration of the complex* $K = K(\mathbf{x}, M)$

Write K as the direct sum of its components $K_p = K_p(\mathbf{x}, M)$. For every $i \in \mathbf{Z}$, define the submodule $F^i K_p$ of K_p by:

$$F^i K_p = \mathbf{x}^{i-p} K_p, \quad \text{where } \mathbf{x}^j = A \text{ if } j \leq 0.$$

The direct sum $F^i K$ of the $F^i K_p$ is a subcomplex of K. We thus get a decreasing filtration:

$$K = F^0 K \supset F^1 K \supset \cdots$$

which is an $\mathbf{x}$-good filtration of K.

(3.2)   Let $\mathrm{gr}(A)$ be the graded ring associated with the $\mathbf{x}$-adic filtration of A. We have $\mathrm{gr}_0(A) = A/\mathbf{x}$ and $\mathrm{gr}_1(A) = \mathbf{x}/\mathbf{x}^2$; denote by $\xi_1, \dots, \xi_r$ the images of $x_1, \dots, x_r$ in $\mathrm{gr}_1(A)$ and put $\xi = (\xi_1, \dots, \xi_r)$. Let $\mathrm{gr}(M)$ be the graded module associated with the $\mathbf{x}$-adic filtration of M; it is a graded $\mathrm{gr}(A)$-module. Then:

The graded complex $gr(K) = \bigoplus_i F^i K / F^{i+1} K$ is isomorphic to the Koszul complex $K(\xi, gr(M))$.

This is clear.

(3.3)   The homology modules $H_p(\xi, gr(M))$ have finite length.

Indeed, they are finitely generated modules over $gr(A)/\xi = A/\mathbf{x}$, and they are annihilated by the ideal $\mathrm{Ann}(M)$. By assumption, the ring $A/(\mathbf{x} + \mathrm{Ann}(M))$ is artinian.

(3.4)   There exists $m \geq 0$ such that $H_p(F^i K / F^{i+1} K) = 0$ for all $i > m$ and all p.

This follows from (3.3) since $H_p(\xi, gr(M))$ is the direct sum of the $H_p(F^i K / F^{i+1} K)$.

Let us now choose m as in (3.4), with $m \geq r$.

(3.5)   We have $H_p(F^i K / F^{i+j} K) = 0$ for $p \in \mathbf{Z}$, $i > m$, $j \geq 0$.

This follows from (3.4) by induction on j, using the exact sequence of complexes:

$$0 \to F^{i+j-1} K / F^{i+j} K \to F^i K / F^{i+j} K \to F^i K / F^{i+j-1} K \to 0.$$

(3.6)   We have $H_p(F^i K) = 0$ for $i > m$, and all p.

Let Z_p^i (resp. B_p^i) denote the module of cycles (resp. boundaries) in $F^i K_p$. By (3.5) we have $H_p(F^i K / \mathbf{x}^j F^i K) = 0$ for every $j \geq 0$. Hence:

$$Z_p^i \subset B_p^i + \mathbf{x}^j F^i K_p \qquad \text{for all } j \geq 0.$$

This means that Z_p^i is contained in the closure of B_p^i for the $\mathbf{x}$-adic topology of $F^i K_p$, hence in B_p^i (cf. Chap. II, cor. 4 to theorem 1). This shows that $H_p(F^i K) = 0$.

(3.7)   If $i > m$, the natural map $H_p(\mathbf{x}, M) = H_p(K) \to H_p(K/F^i K)$ is an isomorphism for every p.

This is a reformulation of (3.6).

(3.8)   We have $\chi(\mathbf{x}, M) = \chi(K/F^i K)$ if $i > m$.

By (3.7), we have

$$\chi(\mathbf{x}, M) = \sum_p (-1)^p \ell(H_p(K/F^i K)).$$

Note that $K/F^i K$ has finite length. It is well-known that, if a complex has finite length, its Euler-Poincaré characteristic is the same as that of its homology. Hence:

$$\sum_p (-1)^p \ell(H_p(K/F^i K)) = \sum_p (-1)^p \ell(K_p/F^i K_p) = \chi(K/F^i K),$$

hence $\chi(\mathbf{x}, M) = \chi(K/F^i K)$.

(3.9)   We have $\chi(\mathbf{x}, M) = \sum_p (-1)^p \binom{r}{p} \ell(M/\mathbf{x}^{i-p} M)$ if $i > m$.

This follows from (3.8) since $K_p/F^i K_p$ is isomorphic to a direct sum of $\binom{r}{p}$ copies of $M/\mathbf{x}^{i-p} M$.

(3.10)   *End of proof*
    If i is large enough, we may rewrite (3.9) as

$$\chi(\mathbf{x}, M) = \sum_p (-1)^p \binom{r}{p} P_\mathbf{x}(M, i - p).$$

A simple computation shows that the right side is equal to $\Delta^r P_\mathbf{x}(M)$, qed.

For future use, let us record:

Theorem 1'.   If $H_p(\xi, \mathrm{gr}(M)) = 0$ for all $p > 0$, then $H_p(\mathbf{x}, M) = 0$ for all $p > 0$.

Indeed, the same argument as above shows that $H_p(F^i K/F^{i+1} K) = 0$ for all $p > 0$ and all i; hence also $H_p(F^i K/F^{i+j} K) = 0$ for $p > 0$, $i, j \geq 0$. As in (3.6), this implies $H_p(F^i K) = 0$ for $p > 0$ and all $i \geq 0$; by choosing $i = 0$, one gets $H_p(K) = 0$ for $p > 0$, qed.

Remark.   The proofs of theorem 1 and theorem 1' could be somewhat shortened by using the spectral sequence associated to the filtration $(F^i K)$, and showing (as in (3.6)) that it is convergent. This method was the one used in the original French text.

4. The depth of a module over a noetherian local ring

In this section, A is a noetherian local ring, with maximal ideal $\mathfrak{m}$ and residue field $k = A/\mathfrak{m}$. All A-modules are assumed to be finitely generated.

Let M be such a module. An M-**sequence** is defined as a sequence $\mathbf{a} = \{a_1, \ldots, a_p\}$ of elements of $\mathfrak{m}$ which satisfy the equivalent three conditions:

a) For every i, $1 \leq i \leq p$, a_i is not a zero-divisor in $M/\mathbf{a}_{i-1}M$, where $\mathbf{a}_0 = 0$ and $\mathbf{a}_{i-1} = (a_1, \dots, a_{i-1})$.

b) $K(\mathbf{a}, M)$ is an acyclic complex (in dimension > 0).

c) $H_1(\mathbf{a}, M) = 0$.

The equivalence of these conditions has already been proved (proposition 3); in particular, these conditions do not depend on the order of the sequence. If we let M_i denote the module $M/\mathbf{a}_iM$, and if $\mathbf{b} = \{b_1, \dots, b_e\}$ is an M_p-sequence, the sequence $\{\mathbf{a}, \mathbf{b}\} = \{a_1, \dots, a_p, b_1, \dots, b_e\}$ is an M-sequence.

Such a sequence $\mathbf{b}$ exists (and has at least one element) if and only if $\mathfrak{m}$ contains an element which is not a zero-divisor in M_p, which is to say if and only if $\mathfrak{m}$ is not associated to M_p.

This last condition is equivalent to the equality $\mathrm{Hom}^A(k, M_p) = 0$, and it depends only on the number p and not on the sequence $\mathbf{a}$, as follows from:

Proposition 5.   *Under the above hypotheses, $\mathrm{Ext}_A^p(k, M)$ is isomorphic to $\mathrm{Hom}(k, M_p)$ and $\mathrm{Ext}_A^i(k, M)$ is 0 for $i < p$.*

The proposition is true if $p = 0$. Thus suppose it has been proved for every A-module N and every N-sequence of fewer than p elements, and let us show it in our case. Since $\{a_2, \dots, a_p\}$ is an M_1-sequence, we have $\mathrm{Hom}(k, M_p) \cong \mathrm{Ext}^{p-1}(k, M_1)$, and it remains to show that $\mathrm{Ext}^{p-1}(k, M_1) \cong \mathrm{Ext}^p(k, M)$. The endomorphism of M defined by a_1 gives rise to the exact sequences:

$$0 \to M \xrightarrow{a_1} M \to M_1 \to 0$$

and

$$\cdots \to \mathrm{Ext}^{p-1}(k, M) \to \mathrm{Ext}^{p-1}(k, M_1) \to \mathrm{Ext}^p(k, M) \xrightarrow{a_1} \mathrm{Ext}^p(k, M).$$

But $\mathrm{Ext}^{p-1}(k, M) = \mathrm{Hom}(k, M_{p-1}) = 0$ and the annihilator of $\mathrm{Ext}^p(k, M)$ contains $\mathrm{Ann}(k) = \mathfrak{m}$ and hence a_1; the homomorphism from $\mathrm{Ext}^{p-1}(k, M_1)$ into $\mathrm{Ext}^p(k, M)$ is therefore an isomorphism. A similar argument shows that $\mathrm{Ext}^i(k, M) = 0$ for $i < p$.

Now suppose $M \neq 0$. Any M-sequence can be embedded in a *maximal* one: if not there would exist an infinite sequence $(a_1, a_2, \dots)$ having property a) above, and the corresponding sequence of ideals $\mathfrak{a}_1 \subset \mathfrak{a}_2 \subset \dots$ would be strictly increasing.

Hence there exists a *maximal M-sequence* $(a_1, \dots, a_p)$; by proposition 5, we have

$$\mathrm{Ext}_A^i(k, M) = 0 \quad \text{for } i < p,$$

and $$\mathrm{Ext}_A^p(k, M) = \mathrm{Hom}_A(k, M_p) \neq 0.$$

This shows that the length p of the sequence depends only on M. In other words:

Proposition and Definition 6.   *All the maximal M-sequences have the same number of elements, say p. Every M-sequence can be extended to a maximal M-sequence. The integer p is the infimum of the n such that $\mathrm{Ext}_A^n(k, M) \neq 0$; it is called the* **depth** *of M and is denoted by* $\mathrm{depth}_A M$.

(In the original French text, the depth was called "codimension homologique"; this terminology is not used any more.)

Corollary .   *With the above notation, one has*

$$\mathrm{depth}_A M_i = \mathrm{depth}_A M - i.$$

Remark.   It is sometimes convenient to extend definition 6 to the trivial module 0 by putting $\mathrm{depth}_A 0 = +\infty$.

Let us again assume that $M \neq 0$.

Proposition 7.
(i) *Every M-sequence can be extended to a system of parameters of M.*
(ii) *One has $\mathrm{depth}_A M \leq \dim A/\mathfrak{p}$ for every $\mathfrak{p} \in \mathrm{Ass}(M)$.*

The proof is by induction on $p = \mathrm{depth}_A M$.

Let $(a_1, \ldots, a_p)$ be a maximal M-sequence. We have $a_1 \notin \mathfrak{p}$ for every $\mathfrak{p} \in \mathrm{Ass}(M)$, cf. Chap. I, prop. 7, hence

$$\dim M/a_1 M = \dim M - 1,$$

cf. Chap. III, cor. 5 to th. 1. By induction on i, we see that

$$\dim(M/(a_1, \ldots, a_i)M) = \dim M - i \qquad \text{for } i = 1, \ldots, p,$$

which implies (i), cf. Chap. III, prop. 6.

Let now $\mathfrak{p}$ be an element of $\mathrm{Ass}(M)$. We have an exact sequence

$$0 \to \mathrm{Hom}(A/\mathfrak{p}, M) \xrightarrow{a_1} \mathrm{Hom}(A/\mathfrak{p}, M) \to \mathrm{Hom}(A/\mathfrak{p}, M/a_1 M).$$

Since $\mathfrak{p}$ belongs to $\mathrm{Ass}(M)$, $\mathrm{Hom}(A/\mathfrak{p}, M)$ is nonzero. By Nakayama's lemma, the same is true for $\mathrm{Hom}(A/\mathfrak{p}, M)/a_1 \mathrm{Hom}(A/\mathfrak{p}, M)$, hence also for $\mathrm{Hom}(A/\mathfrak{p}, M/a_1 M)$. This means that there is a nonzero element of $M/a_1 M$ which is annihilated by $\mathfrak{p} + a_1 A$. By Chap. I, proposition 7, $\mathfrak{p} + a_1 A$ is contained in a prime ideal $\mathfrak{q}$ belonging to $\mathrm{Ass}(M/a_1 M)$. The

induction hypothesis shows that $\dim A/\mathfrak{q} \geq p - 1$. But one has
$$\dim A/\mathfrak{q} = \dim A/\mathfrak{p} - 1,$$
since $A/\mathfrak{q} = (A/\mathfrak{p})/a_1(A/\mathfrak{p})$, and $a_1 \notin \mathfrak{p}$. Hence $\dim A/\mathfrak{p} \geq p$, qed.

Proposition 8.   (cf. [Bour], Chap. X, §1, prop. 1)   *Let*
$$0 \to M' \to M \to M'' \to 0$$
be an exact sequence of A-modules. Define:
$$p = \operatorname{depth}_A M, \quad p' = \operatorname{depth}_A M', \quad p'' = \operatorname{depth}_A M''.$$
Then one has either $p' = p \leq p''$, or $p = p'' < p'$, or $p'' = p' - 1 < p$.

This follows from the exact sequence
$$\ldots \to \operatorname{Ext}_A^{i-1}(k, M'') \to \operatorname{Ext}_A^i(k, M') \to \operatorname{Ext}_A^i(k, M) \to \operatorname{Ext}_A^i(k, M'') \to$$
by considering the first term (on the left) which is $\neq 0$, and noticing that the next one is nonzero too.

Proposition 9.   *Let $\hat{A}$ be the completion of A for the $\mathfrak{m}$-adic topology, and $\hat{M} = \hat{A} \otimes_A M$. Then:*
i) $\operatorname{depth}_{\hat{A}} \hat{M} = \operatorname{depth}_A M$.
ii) *Every maximal M-sequence is a maximal $\hat{M}$-sequence.*

Assertion i) follows from the fact that $\hat{A} \otimes_A \operatorname{Ext}_A^i(k, M)$ is isomorphic to $\operatorname{Ext}_{\hat{A}}^i(k, \hat{M})$.

To prove b), consider a maximal M-sequence $\mathbf{a} = \{a_1, \ldots, a_p\}$. With the same notation as above, we have the exact sequences
$$0 \to M_{i-1} \xrightarrow{a_i} M_{i-1} \to M_i \to 0$$
and therefore also
$$0 \to \hat{M}_{i-1} \xrightarrow{a_i} \hat{M}_{i-1} \to \hat{M}_i \to 0.$$
Hence $\mathbf{a}$ is an $\hat{M}$-sequence, which is maximal because of i).

B: Cohen-Macaulay Modules

(See [Bour], Chap. X, §2 and also [H2], [BrH].)

Let A be a noetherian local ring, with maximal ideal $\mathfrak{m} = \mathfrak{m}(A)$; let E be a nonzero finitely generated A-module.

1. Definition of Cohen-Macaulay modules

By prop. 7, for every $\mathfrak{p} \in \mathrm{Ass}(E)$, we have

$$\dim(A/\mathfrak{p}) \geq \mathrm{depth}(E).$$

Since $\dim E = \sup \dim(A/\mathfrak{p})$ for $\mathfrak{p} \in \mathrm{Ass}(E)$, we have in particular $\dim E \geq \mathrm{depth}\, E$.

Definition 1. The module E is called a **Cohen-Macaulay module** if

$$\dim(E) = \mathrm{depth}(E).$$

The ring A is called a **Cohen-Macaulay ring** if it is a Cohen-Macaulay module when viewed as a module over itself.

Examples.

1) An artinian local ring, a local domain of dimension 1, are Cohen-Macaulay rings.

2) A local integrally closed domain of dimension 2 is a Cohen-Macaulay ring. Indeed, if x is a nonzero element of $\mathfrak{m}$, the prime ideals $\mathfrak{p}$ of $\mathrm{Ass}(A/xA)$ are of height 1, thus distinct from $\mathfrak{m}$ since $\dim A = 2$. We thus conclude that $\mathrm{depth}(A/xA) \geq 1$. whence $\mathrm{depth}(A) \geq 2$, which shows that A is a Cohen-Macaulay ring.

Proposition 10. *For E to be a Cohen-Macaulay module, it is necessary and sufficient that the $\hat{A}$-module $\hat{E}$ is a Cohen-Macaulay module.*

This follows from the formulae

$$\mathrm{depth}(E) = \mathrm{depth}(\hat{E}) \quad \text{and} \quad \dim(E) = \dim(\hat{E}).$$

Proposition 11. *Let A and B be two noetherian local rings and let $\phi : A \to B$ be a homomorphism which makes B into a finitely generated A-module. If E is a finitely generated B-module, then E is a Cohen-Macaulay A-module if and only if it is a Cohen-Macaulay B-module.*

This follows from the following more general proposition:

Proposition 12. *Let A and B be two noetherian local rings, and let $\phi : A \to B$ be a homomorphism which makes B into a finitely generated A-module. If E is a finitely generated B-module, then:*

$$\mathrm{depth}_A(E) = \mathrm{depth}_B(E) \quad \text{and} \quad \dim_A(E) = \dim_B(E).$$

The homomorphism $\phi : A \to B$ maps $\mathfrak{m}(A)$ into $\mathfrak{m}(B)$: if not, we would have $\mathfrak{m}(A)B = B$, contrary to Nakayama's lemma. Write E_A (resp. E_B) for E viewed as an A-module (resp. a B-module). Let $(a_1, \ldots, a_n)$ be a maximal E_A-sequence. If we let $b_i = \phi(a_i)$, the b_i form an E_B-sequence. Furthermore, this E_B-sequence is maximal; indeed, since (a_i) is maximal, there exists a nonzero A-submodule F' of $F = E/(a_1, \ldots, a_n)E$ which is annihilated by $\mathfrak{m}(A)$, and F' generates a B-submodule of F which is of finite length over B, which shows that $(b_1, \ldots, b_n)$ is a maximal E_B-sequence. Thus

$$\text{depth}_A(E_A) = n = \text{depth}_B(E_B).$$

The formula for the dimension is easy (e.g. use prop. 3 of Chap. III).

2. Several characterizations of Cohen-Macaulay modules

Proposition 13.   Let E be a Cohen-Macaulay A-module of dimension n. For every $\mathfrak{p} \in \text{Ass}(E)$, we have $\dim A/\mathfrak{p} = n$, and $\mathfrak{p}$ is a minimal element of $\text{Supp}(E)$.

Indeed, we have $\dim(E) \geq \dim(A/\mathfrak{p}) \geq \text{depth}(E)$ (cf. prop. 7), whence $\dim(A/\mathfrak{p}) = \dim(E) = n$, since the extreme terms are equal. Furthermore, $\mathfrak{p}$ contains a minimal element $\mathfrak{p}'$ of $\text{Supp}(E)$; by th. 1 of Chap. I, we have $\mathfrak{p}' \in \text{Ass}(E)$; the above shows that

$$\dim(A/\mathfrak{p}') = n = \dim(A/\mathfrak{p}),$$

whence $\mathfrak{p}' = \mathfrak{p}$, qed.

Proposition 14.   Let E be a Cohen-Macaulay A-module of dimension $n \geq 1$, and let $x \in \mathfrak{m}$ such that $\dim(E/xE) = n - 1$. Then the endomorphism of E defined by x is injective, and E/xE is a Cohen-Macaulay module.

Let $\mathfrak{p}_1, \ldots, \mathfrak{p}_k$ be the elements of $\text{Ass}(E)$. If x belongs to one of the $\mathfrak{p}_i$, say $\mathfrak{p}_1$, we will have $\mathfrak{p}_1 \in \text{Supp}(E/xE)$, whence $\dim(E/xE) \geq n$. Thus x does not belong to any $\mathfrak{p}_i$, which means (cf. Chap. I, prop. 7) that the endomorphism of E defined by x is injective. It follows that $\text{depth}(E/xE) = \text{depth}(E) - 1$ (cor. to prop. 6), whence the fact that E/xE is Cohen-Macaulay.

Theorem 2.   If E is a Cohen-Macaulay module, every system of parameters of E is an E-sequence. Conversely, if a system of parameters of E is an E-sequence, E is a Cohen-Macaulay module.

Suppose that E is a Cohen-Macaulay module of dimension n; let $(x_1, \ldots, x_n)$ be a system of parameters of E. We will show by induction on k that $(x_1, \ldots, x_k)$ is an E-sequence and that $E/(x_1, \ldots, x_k)E$ is a Cohen-Macaulay module. For $k = 0$, this is clear. We pass from k to $k + 1$ using prop. 14, and observing that $\dim(E/(x_1, \ldots, x_k)E) = n - k$ since the x_i form a system of parameters of E.

The converse is trivial.

Corollary .   If E is a Cohen-Macaulay module, and if $\mathfrak{a}$ is an ideal of A generated by a subset of k elements of a system of parameters of A, the module $E/\mathfrak{a}E$ is a Cohen-Macaulay module of dimension equal to $\dim(E) - k$.

This has been proved along the way.

The condition of theorem 2 can be transformed using the results of part A. Let E be an A-module of dimension n, and let $\mathbf{x} = (x_1, \ldots, x_n)$ be a system of parameters of E; let us also write $\mathbf{x}$ for the ideal generated by the x_i. Let $e_{\mathbf{x}}(E, n)$ denote the multiplicity of $\mathbf{x}$ with respect to E (cf. th. 1), let $H_q(\mathbf{x}, E)$ denote the q-th homology group of the Koszul complex defined by $\mathbf{x}$ and E, and let $\mathrm{gr}_{\mathbf{x}}(E)$ denote the graded module associated to E filtered by the $\mathbf{x}$-adic filtration.

Theorem 3.   Let E be of dimension n. If E is a Cohen-Macaulay module, then for every system of parameters $\mathbf{x} = (x_1, \ldots, x_n)$ of E, we have the following properties:
  i) $e_{\mathbf{x}}(E, n) = \ell(E/\mathbf{x}E)$, length of $E/\mathbf{x}E$.
 ii) $\mathrm{gr}_{\mathbf{x}}(E) = (E/\mathbf{x}E)[X_1, \ldots, X_n]$.
iii) $H_1(\mathbf{x}, E) = 0$.
 iv) $H_q(\mathbf{x}, E) = 0$ for all $q \geq 1$.
Conversely, if a system of parameters of E satisfies any one of these properties, it satisfies all of them and E is a Cohen-Macaulay module.

Each of the properties i), ii), iii), iv) is equivalent to the fact that $\mathbf{x}$ is an E-sequence: for iii) and iv), this is proposition 3; moreover i) and ii) are equivalent (Chap. II, th. 2′); iv) implies i) according to theorem 1; finally ii) implies that the $H_i(\xi, \mathrm{gr}(E))$ are zero for $i \geq 1$, which implies (cf. part A, th. 1′) that $H_i(\mathbf{x}, E) = 0$ for $i \geq 1$. The theorem follows from that.

3. The support of a Cohen-Macaulay module

Theorem 4.   *Let E be a Cohen-Macaulay module of dimension n, and let $x_1,\ldots,x_r \in \mathfrak{m}$ be such that $\dim E/(x_1,\ldots,x_r)E = n-r$. Then every element $\mathfrak{p}$ of $\mathrm{Ass}(E/(x_1,\ldots,x_r)E)$ is such that $\dim(A/\mathfrak{p}) = n-r$.*

The hypothesis means that $x_1,\ldots,x_r$ is a subset of a system of parameters of E. According to the corollary to theorem 1, the quotient module $E/(x_1,\ldots,x_r)E$ is a Cohen-Macaulay module of dimension $n-r$, and the theorem follows by applying prop. 13.

Theorem 4 *characterizes* the Cohen-Macaulay modules. More precisely:

Theorem 5.   *Let E be of dimension n. Suppose that, for every family $(x_1,\ldots,x_r)$ of elements of $\mathfrak{m}$ such that $\dim E/(x_1,\ldots,x_r)E = n-r$, and for every $\mathfrak{p} \in \mathrm{Ass}(E/(x_1,\ldots,x_r)E)$, we have $\dim(A/\mathfrak{p}) = n-r$. Then E is a Cohen-Macaulay module.*

We argue by induction on n, the case $n = 0$ being trivial. Thus suppose $n \geq 1$. Applying the hypotheses to an empty family of elements x_i, we see that $\dim(A/\mathfrak{p}) = n$ for every $\mathfrak{p} \in \mathrm{Ass}(E)$; as $\dim(E) \geq 1$, there is therefore some $x_1 \in \mathfrak{m}(A)$ which belongs to none of the $\mathfrak{p} \in \mathrm{Ass}(E)$. The endomorphism of E defined by x_1 is then injective, and we have:

$$\mathrm{depth}(E) = \mathrm{depth}(E/x_1E) + 1, \quad \dim(E) = \dim(E/x_1E) + 1.$$

Moreover, it is clear that the module E/x_1E satisfies the hypotheses of th. 5 with $n-1$ instead of n; according to the induction hypothesis, it is thus a Cohen-Macaulay module, and so is E.

Theorem 6.   *Let E be a Cohen-Macaulay module of dimension n, and let $\mathfrak{p} \in \mathrm{Supp}(E)$. Then there exists an integer r, and a subset of r elements $x_1,\ldots,x_r$ of a system of parameters of E, such that $\mathfrak{p}$ belongs to $\mathrm{Ass}(E/(x_1,\ldots,x_r)E)$. Further,*

$$\dim(A/\mathfrak{p}) = n-r, \quad \dim(E_\mathfrak{p}) = r,$$

and $E_\mathfrak{p}$ is a Cohen-Macaulay $A_\mathfrak{p}$-module.

Let $x_1,\ldots,x_r$ be a subset of a system of parameters of E contained in $\mathfrak{p}$ and maximal with respect to this property. Let $\mathfrak{p}_i$ be the elements of $\mathrm{Ass}(E/(x_1,\ldots,x_r)E)$; according to th. 4, we have

$$\dim(A/\mathfrak{p}_i) = n-r \quad \text{for every } i.$$

It follows in particular that the $\mathfrak{p}_i$ are the minimal elements of $\text{Supp}(E/(x_1,\dots,x_r)E)$. Since $\mathfrak{p} \in \text{Supp}(E)$, and $x_1,\dots,x_r$ are contained in $\mathfrak{p}$, we have $\mathfrak{p} \in \text{Supp}(E/(x_1,\dots,x_r)E)$, and $\mathfrak{p}$ contains one of the $\mathfrak{p}_i$, say $\mathfrak{p}_1$. I claim that $\mathfrak{p} = \mathfrak{p}_1$. If not, we would have $\dim(A/\mathfrak{p}) < \dim(A/\mathfrak{p}_1) = \dim(A/\mathfrak{p}_i)$, whence $\mathfrak{p} \neq \mathfrak{p}_i$ for every i, and we could find an element x_{r+1} in $\mathfrak{p}$ which belongs to none of the $\mathfrak{p}_i$; the set $x_1,\dots,x_{r+1}$ would thus be a subset of a system of parameters of E, contrary to the maximality of $x_1,\dots,x_r$.

Thus $\mathfrak{p} = \mathfrak{p}_1$, which shows that the x_i satisfy the stated condition, and proves at the same time that $\dim(A/\mathfrak{p}) = n-r$. Moreover, the x_i form an $A_\mathfrak{p}$-sequence of $E_\mathfrak{p}$, which is at the same time a system of parameters, since $\mathfrak{p}$ is a minimal element of $\text{Supp}(E/(x_1,\dots,x_r)E)$. This proves that $E_\mathfrak{p}$ is a Cohen-Macaulay module of dimension r, qed.

Corollary 1. *Every localization of a Cohen-Macaulay ring is a Cohen-Macaulay ring.*

Corollary 2. *Let E be a Cohen-Macaulay module, and let $\mathfrak{p}, \mathfrak{p}'$ be two elements of $\text{Supp}(E)$, with $\mathfrak{p} \subset \mathfrak{p}'$. Then all the saturated chains of prime ideals joining $\mathfrak{p}$ to $\mathfrak{p}'$ have the same length, which is*

$$\dim(A/\mathfrak{p}) - \dim(A/\mathfrak{p}').$$

It suffices to consider the case where $\mathfrak{p}$ and $\mathfrak{p}'$ are consecutive, i.e. where $\dim A_{\mathfrak{p}'}/\mathfrak{p}A_{\mathfrak{p}'} = 1$; then we have to show that

$$\dim(A/\mathfrak{p}) - \dim(A/\mathfrak{p}') = 1.$$

But, applying th. 5 to the module $E_{\mathfrak{p}'}$, we find:

$$\dim E_\mathfrak{p} = \dim E_{\mathfrak{p}'} - \dim A_{\mathfrak{p}'}/\mathfrak{p}A_{\mathfrak{p}'} = \dim E_{\mathfrak{p}'} - 1.$$

Applying it to E, we find:

$$\dim E_\mathfrak{p} = \dim E - \dim A/\mathfrak{p},$$
$$\dim E_{\mathfrak{p}'} = \dim E - \dim A/\mathfrak{p}'.$$

Eliminating $\dim E_\mathfrak{p}$ and $\dim E_{\mathfrak{p}'}$ from these three equations, we obtain $\dim A/\mathfrak{p} - \dim A/\mathfrak{p}' = 1$, qed.

Corollary 3. *Let A be a quotient of a Cohen-Macaulay ring, and let $\mathfrak{p} \subset \mathfrak{p}'$ be two prime ideals of A. Then all the saturated chains of prime ideals joining $\mathfrak{p}$ to $\mathfrak{p}'$ have the same length, which is $\dim A/\mathfrak{p} - \dim A/\mathfrak{p}'$.*

We reduce to the case of a Cohen-Macaulay ring, which follows from corollary 2.

Corollary 4.   *Let A be a domain, which is a quotient of a Cohen-Macaulay ring, and let $\mathfrak{p}$ be a prime ideal of A. We have*

$$\dim A = \dim A_\mathfrak{p} + \dim A/\mathfrak{p}.$$

This follows from corollary 3.

Remark.   Corollaries 3 and 4 are interesting because of the fact that the local rings of algebraic (and analytic) geometry are quotients of Cohen-Macaulay rings — and in fact are quotients of regular rings, cf. part D.

4. Prime ideals and completion

Let $\hat{A}$ be the completion of A. If $\mathfrak{p}$ is a prime ideal of A, the ideal $\mathfrak{p}\hat{A}$ is not in general prime in $\hat{A}$. It may even happen that its primary decomposition involves *embedded* prime ideals. We propose to show that this unpleasant phenomenon does not occur when A is a Cohen-Macaulay ring.

We first prove a general result:

Proposition 15.   *Let A and B be two noetherian rings, B being an A-algebra. Suppose that B is A-flat. Let E be a finitely generated A-module. Then:*

$$\mathrm{Ass}_B(E \otimes_A B) = \bigcup_{\mathfrak{p} \in \mathrm{Ass}_A(E)} \mathrm{Ass}_B(B/\mathfrak{p}B). \qquad (*)$$

Let $\mathfrak{p} \in \mathrm{Ass}(E)$. We have an exact sequence $0 \to A/\mathfrak{p} \to E$, whence, since B is A-flat, an exact sequence $0 \to B/\mathfrak{p}B \to E \otimes_A B$, and hence

$$\mathrm{Ass}_B(B/\mathfrak{p}B) \subset \mathrm{Ass}_B(E \otimes_A B).$$

Thus we have proved that the right side of the formula $(*)$ is contained in the left side.

For proving the reverse inclusion, let $\mathfrak{p}_1, \ldots, \mathfrak{p}_k$ be the elements of $\mathrm{Ass}(E)$, and let

$$E \to \bigoplus E_i$$

be an embedding, where the A-modules E_i ($i = 1, \ldots, k$) are such that $\mathrm{Ass}(E_i) = \{\mathfrak{p}_i\}$, cf. Chap. I, prop. 10. We have

$$\mathrm{Ass}_B(E \otimes_A B) \subset \bigcup_i \mathrm{Ass}_B(E_i \otimes_A B),$$

and we are reduced to showing that $\mathrm{Ass}_B(E_i \otimes_A B) \subset \mathrm{Ass}_B(B/\mathfrak{p}_i B)$; i.e. we are reduced to the case where $\mathrm{Ass}(E)$ has only one element $\mathfrak{p}$.

Let us consider this case. By th. 1 of Chap. I, there is a composition series of E which consists of modules of the type $A/\mathfrak{q}_\alpha$, where $\mathfrak{q}_\alpha$ is a prime ideal containing $\mathfrak{p}$. Passing to $E \otimes_A B$, we conclude that:

$$\mathrm{Ass}_B(E \otimes_A B) \subset \mathrm{Ass}_B(B/\mathfrak{p}B) \cup \bigcup_\alpha \mathrm{Ass}_B(B/\mathfrak{q}_\alpha B),$$

where the $\mathfrak{q}_\alpha$ strictly contain $\mathfrak{p}$. Let $S = A - \mathfrak{p}$; the endomorphisms defined by the elements of S are injective on E, and therefore also on $E \otimes_A B$ since B is A-flat; thus $\mathfrak{p}' \cap S = \emptyset$ for every $\mathfrak{p}' \in \mathrm{Ass}(E \otimes_A B)$. Moreover, since $\mathfrak{q}_\alpha$ strictly contains $\mathfrak{p}$, we have $(A/\mathfrak{q}_\alpha)_S = 0$, hence $(B/\mathfrak{q}_\alpha B)_S = 0$, and $\mathfrak{p}' \cap S \neq \emptyset$ for every $\mathfrak{p}' \in \mathrm{Ass}_B(B/\mathfrak{q}_\alpha B)$. Thus $\mathrm{Ass}_B(E \otimes_A B) \cap \mathrm{Ass}_B(B/\mathfrak{q}_\alpha B) = \emptyset$, which concludes the proof.

Theorem 7.   *Suppose that A is a Cohen-Macaulay local ring, and let $\mathfrak{p}$ be a prime ideal of A. Then every element $\cdot\mathfrak{p}' \in \mathrm{Ass}_{\hat{A}}(\hat{A}/\mathfrak{p}\hat{A})$ is such that* $\dim \hat{A}/\mathfrak{p}' = \dim A/\mathfrak{p}$ *(the ideal $\mathfrak{p}\hat{A}$ thus has no embedded component).*

Let $r = \dim A - \dim A/\mathfrak{p}$. According to theorem 6, there exists a subset of r elements $x_1, \ldots, x_r$ of a system of parameters of A such that $\mathfrak{p} \in \mathrm{Ass}(E)$, where $E = A/(x_1, \ldots, x_r)A$. Moreover, according to theorem 4, the module E is a Cohen-Macaulay module of dimension $\dim A/\mathfrak{p}$. The same is thus true for its completion $\hat{E}$. According to proposition 15, (which is applicable since $\hat{A}$ is A-flat), we have $\mathrm{Ass}(\hat{A}/\mathfrak{p}\hat{A}) \subset \mathrm{Ass}(\hat{E})$. But, according to proposition 13 applied to $\hat{E}$, every $\mathfrak{p}' \in \mathrm{Ass}(\hat{E})$ is such that $\dim \hat{A}/\mathfrak{p}' = \dim \hat{E}$, whence the result.

Corollary .   *Let E be a finitely generated module over a Cohen-Macaulay local ring, and let n be an integer ≥ 0. If every $\mathfrak{p} \in \mathrm{Ass}(E)$ is such that $\dim A/\mathfrak{p} = n$, then the same is true for every $\mathfrak{p}' \in \mathrm{Ass}(\hat{E})$.*

This follows from th. 7, combined with prop. 15.

Remark.   It would be even more pleasant if we had $\mathfrak{p}\hat{A} = \bigcap \mathfrak{p}'$, in the notation of theorem 7. However, this is false in general, even if A is regular ([Nag3], p. 209, example 7). It is nevertheless true for the local rings of algebraic geometry [Ch2], and more generally for the "excellent" rings of Grothendieck ([EGA], Chap. IV, §7.8); see also [Bour], Chap. IX, §4, th. 3.

C: Homological Dimension and Noetherian Modules

1. The homological dimension of a module

We first recall some definitions from [CaE]. If A is a commutative ring (noetherian or not) and if M is an A-module (finitely generated or not), one defines:

- the **homological or projective dimension of** M as the supremum (finite or infinite) $\operatorname{proj\,dim}_A M$ of the integers p such that $\operatorname{Ext}^p_A(M, N) \neq 0$ for at least one A-module N,
- the **injective dimension of** M as the supremum $\operatorname{inj\,dim}_A M$ of the integers p such that $\operatorname{Ext}^p_A(N, M) \neq 0$ for at least one A-module N,
- the **global homological dimension of** A as the supremum $\operatorname{glob\,dim} A$ of the integers p such that $\operatorname{Ext}^p_A(M, N) \neq 0$ for at least one pair of A-modules.

Saying that $\operatorname{proj\,dim}_A(M) = 0$ (resp. $\operatorname{inj\,dim}_A M = 0$) is the same as saying that M is projective (resp. injective).

The following inequalities are direct consequences of the properties of the bifunctor $(M, N) \mapsto \operatorname{Ext}^p_A(M, N)$:

If the sequence $0 \to M' \to M \to M'' \to 0$ is exact, then:
  i) $\operatorname{proj\,dim}_A M \leq \sup(\operatorname{proj\,dim}_A M', \operatorname{proj\,dim}_A M'')$, and if the inequality is strict, we have $\operatorname{proj\,dim}_A M'' = \operatorname{proj\,dim}_A M' + 1$;
  ii) $\operatorname{inj\,dim}_A M \leq \sup(\operatorname{inj\,dim}_A M', \operatorname{inj\,dim}_A M'')$, and if the inequality is strict, we have $\operatorname{inj\,dim}_A M' = \operatorname{inj\,dim}_A M'' + 1$;
  iii) $\operatorname{proj\,dim}_A M'' \leq \sup(\operatorname{proj\,dim}_A M, \operatorname{proj\,dim}_A M' + 1)$, and if the inequality is strict, we have $\operatorname{proj\,dim}_A M = \operatorname{proj\,dim}_A M'$.

Similarly, if $0 = M_0 \subset M_1 \subset \ldots \subset M_n = M$ is a composition series of M,

$$\operatorname{proj\,dim}_A M \leq \sup_{1 \leq i \leq n} \operatorname{proj\,dim}_A(M_i/M_{i-1}).$$

Proposition 16. *For any A-module M, $\operatorname{inj\,dim}_A M$ is the supremum of the integers p such that $\operatorname{Ext}^p_A(N, M) \neq 0$ for at least one finitely generated A-module N.*

Indeed, let dM be this supremum. We obviously have the inequality $\operatorname{inj\,dim}_A M \geq dM$ and equality holds if $dM = +\infty$. Hence suppose that dM is finite.

If $dM = 0$, $\operatorname{Ext}^1_A(A/\mathfrak{a}, M) = 0$ for every ideal $\mathfrak{a}$ of A and every homomorphism of $\mathfrak{a}$ into M extends to A; whence M is injective ([CaE],

Chap. I) and $\operatorname{inj dim}_A M = dM = 0$.

Suppose now that the result has been proved for $dM < n$ and let us show it for $dM = n$ ($n > 0$). There exists an exact sequence

$$0 \to M \to Q \to N \to 0$$

where Q is an injective module and $\operatorname{inj dim}_A M = \operatorname{inj dim}_A N + 1$. We have $dM = dN + 1$, and $\operatorname{inj dim}_A N = dN$ (induction hypothesis). Whence $dM = \operatorname{inj dim}_A M$.

Corollary (Auslander). $\operatorname{glob dim} A = \sup \operatorname{proj dim}_A M$, where M ranges over all finitely generated A-modules.

Indeed, if we let $d(M, N)$ denote the supremum of the integers p such that $\operatorname{Ext}_A^p(M, N) \neq 0$, we have the equalities:

$$\operatorname{glob dim} A = \sup_{M,N} d(M, N) = \sup_N (\sup_M d(M, N))$$

$$= \sup_N \operatorname{inj dim}_A N$$

$$= \sup_N (\sup_{M'} d(M', N)) = \sup_{M'} (\sup_N d(M', N))$$

$$= \sup_{M'} \operatorname{proj dim}_A M',$$

where M and N range over all A-modules, and M' ranges over all finitely generated A-modules.

2. The noetherian case

From now on, we suppose again that A is a *noetherian ring* and M is a *finitely generated A-module*.

Then $\operatorname{proj dim}_A M$ is the supremum of the integers p such that $\operatorname{Ext}_A^p(M, N) \neq 0$ for at least one finitely generated A-module N ([CaE], Chap. VI, proposition 2.5). Now every such N has a composition series

$$0 = N_0 \subset \ldots \subset N_n = N$$

such that $N_i/N_{i-1} \cong A/\mathfrak{p}_i$, where $\mathfrak{p}_i$ is a prime ideal of A. It follows, in the notation of the preceding section, that $d(M, N) \leq \sup_i d(M, A/\mathfrak{p}_i)$ and that $\operatorname{proj dim}_A M \leq \sup_{\mathfrak{p}} d(M, A/\mathfrak{p})$ where $\mathfrak{p}$ ranges over the prime ideals of A. Proposition 2.1 of Chap. VI of [CaE] can thus be restated as:

Proposition 17.   Let n be an integer ≥ 0. The following assertions are equivalent:
  a) $\operatorname{proj dim}_A M \leq n$.
  b) $\operatorname{Ext}_A^{n+1}(M, A/\mathfrak{p}) = 0$ for all prime ideals $\mathfrak{p}$ of A.

c) For every exact sequence $0 \to M_n \to \ldots \to M_0 \to M \to 0$ such that $M_0, \ldots, M_{n-1}$ are projective, M_n is projective.

d) There exists an exact sequence $0 \to M_n \to \ldots \to M_0 \to M \to 0$, where the M_i are projective, $0 \le i \le n$.

Of course, the A-modules $\operatorname{Ext}^p_A(M, N)$ and $\operatorname{Tor}^A_p(M, N)$ are finitely generated if M and N are; indeed, if M is finitely generated, there exists an exact sequence:

$$M_n \to \ldots \to M_1 \to M_0 \to M \to 0,$$

where the M_i ($i \ge 0$) are finitely generated free modules; the modules $\operatorname{Ext}^p_A(M, N)$ and $\operatorname{Tor}^A_p(M, N)$ are thus quotients of submodules of $\operatorname{Hom}_A(M_p, N)$ and $M_p \otimes_A N$, and these are obviously finitely generated.

Proposition 18. *Let M and N be A-modules with M finitely generated. If $\phi : A \to B$ is a homomorphism from A into B, and if B is A-flat, then we have natural isomorphisms:*

$$\operatorname{Tor}^A_p(M, N) \otimes_A B \cong \operatorname{Tor}^B_p(M \otimes_A B, N \otimes_A B);$$

$$\operatorname{Ext}^p_A(M, N) \otimes_A B \cong \operatorname{Ext}^p_B(M \otimes_A B, N \otimes_A B).$$

We give the proof for "Ext" (note that the isomorphism for "Tor" holds without the finiteness hypothesis).

If, with the above notation, $\underline{M}$ is the complex defined by $(\underline{M})_n = M_n$ and $d_n = \phi_n$, $M_n \otimes_A B$ is B-free and the complex $\underline{M} \otimes_A B$ gives a projective resolution of $M \otimes_A B$. Thus:

$$\operatorname{Ext}^p_B(M \otimes_A B, N \otimes_A B) \cong H^p(\operatorname{Hom}_B(\underline{M} \otimes_A B, N \otimes_A B))$$

$$\cong H^p(\operatorname{Hom}_A(\underline{M}, N) \otimes_A B)$$

(because M_n is a finitely generated free module).

But B being A-flat, we obviously have:

$$H^p(\operatorname{Hom}_A(\underline{M}, N) \otimes_A B) \cong H^p(\operatorname{Hom}_A(\underline{M}, N)) \otimes_A B$$

$$= \operatorname{Ext}^p_A(M, N) \otimes_A B,$$

qed.

This proposition applies when $B = A[X]$, where X is an indeterminate, when $B = \hat{A}$ is the completion of A for an $\mathfrak{m}$-adic topology, or when $B = A_S$ where S is a multiplicatively closed subset of A.

Corollary 1. *If $(A, \mathfrak{m})$ is a Zariski ring and M is a finitely generated A-module given with the $\mathfrak{m}$-adic filtration, we have:*

$$\operatorname{proj \, dim}_A M = \operatorname{proj \, dim}_{\hat{A}} \hat{M}.$$

Indeed, if $\mathrm{Ext}^n(M,N) \neq 0$, $\mathrm{Ext}^n(M,N)$ is Hausdorff and its completion $\mathrm{Ext}^n(\hat{M},\hat{N})$ is nonzero; whence $\mathrm{proj\,dim}_{\hat{A}}\,\hat{M} \geq \mathrm{proj\,dim}_A M$. The reverse inequality follows from the more general property:

Proposition 19.   *Under the hypotheses of the above proposition,*
$$\mathrm{proj\,dim}_B(B \otimes_A M) \leq \mathrm{proj\,dim}_A M.$$

Indeed, if $0 \to M_n \to \ldots \to M_0 \to M \to 0$ is a projective resolution of M, the sequence
$$0 \to M_n \otimes_A B \to \ldots \to M_0 \otimes_A B \to M \otimes_A B \to 0$$
is a projective resolution of $M \otimes_A B$.

Corollary 2.   *One has*
$$\mathrm{proj\,dim}_A M \;=\; \sup_{\mathfrak{p}} \mathrm{proj\,dim}_{A_{\mathfrak{p}}} M_{\mathfrak{p}} \;=\; \sup_{\mathfrak{m}} \mathrm{proj\,dim}_{A_{\mathfrak{m}}} M_{\mathfrak{m}},$$
where $\mathfrak{p}$ ranges over the prime ideals of A and $\mathfrak{m}$ over the maximal ideals.

Indeed, according to the above proposition, we have
$$\mathrm{proj\,dim}_{A_{\mathfrak{p}}} M_{\mathfrak{p}} \leq \mathrm{proj\,dim}_A M.$$
Moreover, if $\mathrm{Ext}_A^p(M,N) = P \neq 0$, $P_{\mathfrak{m}}$ is different from 0 for at least one maximal ideal $\mathfrak{m}$; whence the assertion.

Corollary 2 reduces the study of the homological dimension to the case of modules over a *local ring*.

3. The local case

Proposition 20.   *If A is a noetherian local ring, $\mathfrak{m}$ its maximal ideal, $k = A/\mathfrak{m}$ its residue field and if M is a finitely generated A-module, the following properties are equivalent:*
a) M *is free.*
b) M *is projective.*
b) M *is flat.*
d) $\mathrm{Tor}_1(M,k) = 0$.

The implications a) $\Rightarrow$ b) $\Rightarrow$ c) $\Rightarrow$ d) are clear and it remains to show that d) $\Rightarrow$ a).
Thus suppose that $\mathrm{Tor}_1(M,k) = 0$ and let $x_1,\ldots,x_n$ be elements of M whose images in $M/\mathfrak{m}M$ form a k-basis. Let P be a free A-module

with basis $e_1, \ldots, e_n$ and ϕ the homomorphism from P into M which maps e_i to x_i. By Nakayama's lemma, ϕ is surjective. Let N be its kernel. The exact sequence

$$0 \to N \to P \to M \to 0$$

gives rise to the exact sequence:

$$\mathrm{Tor}_1(P, k) = 0 \to \mathrm{Tor}_1(M, k) = 0 \to N/\mathfrak{m}N \to P/\mathfrak{m}P \xrightarrow{\overline{\phi}} M/\mathfrak{m}M \to 0.$$

As $\overline{\phi}$ is injective, $N/\mathfrak{m}N$ and hence N are therefore zero, qed.

Corollary .   If A is a noetherian ring and M is a finitely generated A-module, M is projective if and only if for every maximal ideal $\mathfrak{m}$ of A, $M_\mathfrak{m}$ is a free $A_\mathfrak{m}$-module.

This follows from the equality: $\mathrm{proj\,dim}_A M = \sup_\mathfrak{m} \mathrm{proj\,dim}_{A_\mathfrak{m}} M_\mathfrak{m}$.

Theorem 8.   Let n be an integer ≥ 0. Under the hypotheses of the preceding proposition, the following properties are equivalent:
a) $\mathrm{proj\,dim}_A M \leq n$.
b) $\mathrm{Tor}_p^A(M, N) = 0$ for all $p > n$ and for all A-modules N.
c) $\mathrm{Tor}_{n+1}^A(M, k) = 0$.

It is clear that a) $\Rightarrow$ b) $\Rightarrow$ c). Let us show that c) $\Rightarrow$ a). To do so, choose an exact sequence:

$$0 \to M_n \xrightarrow{\phi_n} M_{n-1} \xrightarrow{\phi_{n-1}} \ldots \to M_0 \xrightarrow{\phi_0} M \to 0$$

where the modules $M_0, M_1, \ldots, M_{n-1}$ are free. Define

$$Z_i = \mathrm{Ker}\,\phi_i, \quad 0 \leq i \leq n-1.$$

Then the sequence $0 \to Z_i \to M_i \to Z_{i-1} \to 0$ is exact and

$$\mathrm{Tor}_j(Z_i, k) = \mathrm{Tor}_{j+1}(Z_{i-1}, k) \quad \text{if } j \geq 1.$$

It follows that:

$$\mathrm{Tor}_1(M_n, k) = \mathrm{Tor}_2(Z_{n-2}, k) = \ldots$$
$$= \mathrm{Tor}_n(Z_0, k) = \mathrm{Tor}_{n+1}(M, k) = 0,$$

hence M_n is free and a) is true.

Corollary 1.   If M is a finitely generated module over a noetherian ring A, the following properties are equivalent:
a) $\mathrm{proj\,dim}_A M \leq n$.
b) $\mathrm{Tor}_p^A(M, N) = 0$ for all $p > n$ and for all A-modules N.
c) $\mathrm{Tor}_{n+1}^A(M, A/\mathfrak{m}) = 0$ for every maximal ideal $\mathfrak{m}$.

This follows from the theorem and the two preceding propositions.

Corollary 2.   *If A is a noetherian ring, the following properties are equivalent:*
a) $\operatorname{glob dim} A \leq n$.
b) $\operatorname{Tor}^A_{n+1}(A/\mathfrak{m}, A/\mathfrak{m}) = 0$ *for every maximal ideal* $\mathfrak{m}$.

It is trivial that a) $\Rightarrow$ b). Conversely, if $\operatorname{Tor}^A_{n+1}(A/\mathfrak{m}, A/\mathfrak{m}) = 0$, then $\operatorname{Tor}^A_{n+1}(A/\mathfrak{m}, A/\mathfrak{n})$ is zero for every maximal ideal $\mathfrak{n}$ (the annihilator of $\operatorname{Tor}^A_p(M, N)$ contains the annihilators of M and of N). Thus $\operatorname{proj dim}_A(A/\mathfrak{m}) \leq n$ and there exists a projective resolution

$$0 \to L_n \to \ldots \to L_0 \to A/\mathfrak{m} \to 0.$$

But this implies that $\operatorname{Tor}^A_{n+1}(M, A/\mathfrak{m}) = 0$ for every M; whence a).

D: Regular Rings

Definition .   A **regular ring** is a noetherian ring of finite global homological dimension.

1. Properties and characterizations of regular local rings

Let A be a regular local ring, $n = \operatorname{glob dim} A$, $\mathfrak{m}$ the maximal ideal of A, $k = A/\mathfrak{m}$ and M a nonzero finitely generated A-module. The following proposition compares $\operatorname{proj dim}_A M$ and $\operatorname{depth}_A M$:

Proposition 21. $\operatorname{proj dim}_A M + \operatorname{depth}_A M = n$.

The proposition is true if $\operatorname{depth}_A M = 0$, for then there exists an injection of k into M, and, as Tor_n is left exact, an injection of $\operatorname{Tor}_n(k, k)$ into $\operatorname{Tor}_n(M, k)$:

$$0 \to \operatorname{Tor}_n(k, k) \to \operatorname{Tor}_n(M, k).$$

But $\operatorname{Tor}_n(k, k)$ is nonzero (see corollary 2 to theorem 8) and so is $\operatorname{Tor}_n(M, k)$, whence $\operatorname{proj dim}_A M = n$.

Now suppose that the proposition is proved for every module whose depth is strictly less than $\operatorname{depth}_A M$, and let us prove it for M.

It suffices to consider the case where $\operatorname{depth}_A M > 0$, i.e. where there exists $a \in \mathfrak{m}$ which is not a zero-divisor in M. We have an exact sequence:

$$0 \to M \xrightarrow{a} M \to M_1 \to 0,$$

where $M_1 = M/aM$, and $\operatorname{depth}_A M_1 = \operatorname{depth}_A M - 1$. Since $\operatorname{depth}_A M_1 + \operatorname{proj\,dim}_A M_1 = n$ by induction hypothesis, it remains to prove that $\operatorname{proj\,dim}_A M_1 = \operatorname{proj\,dim}_A M + 1$.

But, in the homology sequence:

$$\operatorname{Tor}_p(M,k) \xrightarrow{a} \operatorname{Tor}_p(M,k) \to \operatorname{Tor}_p(M_1,k) \to$$
$$\operatorname{Tor}_{p-1}(M,k) \xrightarrow{a} \operatorname{Tor}_{p-1}(M,k),$$

a belongs to the annihilator of k; hence we have the exact sequence:

$$0 \to \operatorname{Tor}_p(M,k) \to \operatorname{Tor}_p(M_1,k) \to \operatorname{Tor}_{p-1}(M,k) \to 0.$$

Since $\operatorname{Tor}_{p-1}(M,k) = 0$ implies $\operatorname{Tor}_p(M,k) = 0$, we have the equivalence:

$$\operatorname{Tor}_p(M_1,k) = 0 \iff \operatorname{Tor}_{p-1}(M,k) = 0,$$

qed.

Corollary .   For $\operatorname{proj\,dim}_A M$ to be equal to n, it is necessary and sufficient that $\mathfrak{m}$ is associated to M.

The following theorem (cf. [Se2]) shows that our homological definition of regular rings coincide with the usual one.

Theorem 9.   Let A be a noetherian local ring of dimension r, with maximal ideal $\mathfrak{m}$ and residue field $k = A/\mathfrak{m}$. The following properties are equivalent:

a) A is regular.
b) $\mathfrak{m}$ can be generated by r elements.
c) The dimension over k of the vector space $\mathfrak{m}/\mathfrak{m}^2$ is r.
d) The graded ring $\operatorname{gr}_\mathfrak{m}(A)$, associated to the $\mathfrak{m}$-adic filtration of A, is isomorphic to the polynomial algebra $k[X_1, \dots, X_r]$.

The canonical map from $\mathfrak{m}$ onto $\mathfrak{m}/\mathfrak{m}^2$ gives a surjective correspondence between the minimal systems of generators of $\mathfrak{m}$ and the k-bases of $\mathfrak{m}/\mathfrak{m}^2$. Hence b) $\Leftrightarrow$ c). Moreover, it is clear that d) $\Rightarrow$ c). Conversely, b) $\Rightarrow$ d) because if $\mathfrak{m}$ is generated by r elements we have the inequalities:

$$1 \leq \Delta^r P_\mathfrak{m}(A,n) = e_\mathfrak{m}(A,r) \leq \ell(A/\mathfrak{m}) = 1,$$

whence $e_\mathfrak{m}(A,r) = \ell(A/\mathfrak{m})$ and proposition 9 of Chap. II applies.

Let us show that d) $\Rightarrow$ a). Let $\mathbf{x} = (x_1, \dots, x_r)$ be a minimal system of generators of $\mathfrak{m}$. Property d) implies that $\mathbf{x}$ is an A-sequence (this follows from th. 3 applied to the A-module A). In other words, the complex $K(\mathbf{x}, A)$ gives a free resolution of k :

$$0 \to K_r(\mathbf{x}, A) \xrightarrow{d_r} \ldots \xrightarrow{d_1} K_0(\mathbf{x}, A) \xrightarrow{\epsilon} k \to 0.$$

where $K_0(\mathbf{x}, A) \cong A$ and ϵ is the canonical map from A onto k. Thus for every A-module M, we have the equality $\mathrm{Tor}_i(M, k) \cong H_i(\mathbf{x}, M)$.

In particular, $\mathrm{Tor}_i(k, k) \cong H_i(\mathbf{x}, k) = K_i(\mathbf{x}, k) = K_i(\mathbf{x}) \otimes_A k$, whence $\mathrm{Tor}_i^A(k, k) \cong \bigwedge^i(k^r)$, $\mathrm{Tor}_{r+1}^A(k, k) = 0$, $\mathrm{Tor}_r^A(k, k) = k$. Thus $\mathrm{glob\,dim}\,A = r < +\infty$, qed.

It remains to show that a) $\Rightarrow$ c) : Let $n = \mathrm{glob\,dim}\,A$. From

$$\mathrm{proj\,dim}_A A = 0 \quad \text{and} \quad \mathrm{proj\,dim}_A A + \mathrm{depth}_A A = n$$

we find $\mathrm{depth}_A A = n$, and

$$n = \mathrm{depth}_A A \leq \dim A = r.$$

Now the canonical map from $K_i(\mathbf{x}, k)$ into $\mathrm{Tor}_i(k, k)$ is an *injection*, where $\mathbf{x} = (x_1, \ldots, x_s)$, $s \geq r$, denotes a minimal system of generators of $\mathfrak{m}$, i.e. induces a basis of $\mathfrak{m}/\mathfrak{m}^2$ (this is valid for every local ring A; for a proof see Appendix I); hence $\mathrm{Tor}_s(k, k) \neq 0$ and we have $n \geq s$. Whence the inequalities:

$$r \leq s \leq n = \mathrm{depth}_A A \leq \dim A = r,$$

and the result.

Corollary 1.   *If A is regular, then $\dim A = \mathrm{glob\,dim}\,A$.*

Indeed, the equality $r = n$ has been proved along the way.

Corollary 2.   *Assume A is regular, and let $\mathbf{x} = (x_1, \ldots, x_r)$ be a system of parameters generating $\mathfrak{m}$. Let M be a finitely generated A-module.*

i) For $i \geq 0$, one has natural isomorphisms

$$\mathrm{Tor}_i^A(k, M) \cong H_i(\mathbf{x}, M) \cong \mathrm{Ext}_A^{r-i}(k, M).$$

ii) Assume $\dim M = r$. Then M is Cohen-Macaulay if and only if it is free.

We have seen that the Koszul complex $K(\mathbf{x})$ gives a free resolution of k. This implies i), (cf. also cor. 2 and cor. 3 to prop. 3) and ii) follows by applying prop. 21, together with the fact that $r = n$.

Corollary 3.   *A regular local ring is normal, and Cohen-Macaulay.*

If A is regular, it is Cohen-Macaulay by cor. 2, applied to $M = A$; it is normal, because $\mathrm{gr}_\mathfrak{m}(A)$ is normal, cf. Chap. II, part A, §4.

Corollary 4 (Auslander-Buchsbaum, [AuB3]).   *A regular local ring is factorial.*

This is a general property of normal domains in which every ideal has a finite free resolution (cf. [Bour], Chap. VII, §4, p. 68).

Corollary 5.   *A noetherian local ring of dimension 0 (resp. 1) is regular if and only if it is a field (resp. a discrete valuation ring).*

This is clear.

Corollary 6.   *Assume A is regular of dimension 2. Let M be a finitely generated A-module. The following properties are equivalent:*
  (i) *M is free.*
  (ii) *There exists a finitely generated module N such that M is isomorphic to $\mathrm{Hom}(N, A)$.*
  (iii) *M is **reflexive**, i.e. the canonical map $M \rightarrow \mathrm{Hom}(\mathrm{Hom}(M, A), A)$ is an isomorphism.*

The implications (i) $\Rightarrow$ (iii) $\Rightarrow$ (ii) are clear. Let us show (ii) $\Rightarrow$ (i). Choose N as in (ii), and write it as L/R where L is free and finitely generated. We have an exact sequence

$$0 \rightarrow M \rightarrow L' \rightarrow X \rightarrow 0,$$

where $L' = \mathrm{Hom}(L, A)$ and X is the image of L' in $R' = \mathrm{Hom}(R, A)$. Since L' is free, we have $\mathrm{depth}_A L' = 2$. On the other hand, neither M nor X contain a submodule isomorphic to k. Hence we have

$$\mathrm{depth}_A M \geq 1 \quad \text{and} \quad \mathrm{depth}_A X \geq 1.$$

Proposition 8 shows that $\mathrm{depth}_A M \neq 1$. Hence $\mathrm{depth}_A M \geq 2$ and M is a Cohen-Macaulay module, hence is free by corollary 2 above.

(This applies in particular when A is the *Iwasawa algebra* $\mathbf{Z}_p[[T]]$.)

2. Permanence properties of regular local rings

If A is a regular local ring, a **regular system of parameters of A** is defined as any system $\mathbf{x} = \{x_1, \ldots, x_n\}$ of parameters of A which generates the maximal ideal m. We already know that every system of parameters of A is an A-sequence. Among such systems, the regular systems are characterized by:

Proposition 22. If $\{x_1, \ldots, x_p\}$ are p elements of the maximal ideal $\mathfrak{m}$ of a regular local ring A, the following three properties are equivalent:
a) $x_1, \ldots, x_p$ is a subset of a regular system of parameters of A.
b) The images of $x_1, \ldots, x_p$ in $\mathfrak{m}/\mathfrak{m}^2$ are linearly independent over k.
c) The local ring $A/(x_1, \ldots, x_p)$ is regular, and has dimension $\dim A - p$.
(In particular, $(x_1, \ldots, x_p)$ is a prime ideal.)

a) $\Leftrightarrow$ b): Indeed the regular systems of parameters of A correspond to k-bases of $\mathfrak{m}/\mathfrak{m}^2$.

a), b) $\Rightarrow$ c): Indeed we have an exact sequence:

$$0 \to \mathfrak{p}/\mathfrak{p} \cap \mathfrak{m}^2 \to \mathfrak{m}/\mathfrak{m}^2 \to \mathfrak{n}/\mathfrak{n}^2 \to 0$$

where $\mathfrak{p} = (x_1, \ldots, x_p)$ and $\mathfrak{n} = \mathfrak{m}/\mathfrak{p}$ and hence the equivalences:

b) $\Longleftrightarrow$ $[\mathfrak{p}/\mathfrak{p} \cap \mathfrak{m}^2 : k] = p \Longleftrightarrow [\mathfrak{n}/\mathfrak{n}^2 : k] = \dim A - p$.

But $x_1, \ldots, x_p$ form a subset of a system of parameters of A, so $A/(x_1, \ldots, x_p)$ has dimension $\dim A - p$; whence the result.

c) $\Rightarrow$ b): Indeed, c) is equivalent to the two conditions:

$$[\mathfrak{n}/\mathfrak{n}^2 : k] = \dim A/\mathfrak{p} \quad \text{and} \quad \dim A/\mathfrak{p} = \dim A - p.$$

Corollary. If $\mathfrak{p}$ is an ideal of a regular local ring A, the following two properties are equivalent:
a) $A/\mathfrak{p}$ is a regular local ring.
b) $\mathfrak{p}$ is generated by a subset of a regular system of parameters of A.

Only the implication a) $\Rightarrow$ b) remains to be proved. But if $\mathfrak{n} = \mathfrak{m}/\mathfrak{p}$, we have the exact sequence:

$$0 \to \mathfrak{p}/\mathfrak{p} \cap \mathfrak{m}^2 \to \mathfrak{m}/\mathfrak{m}^2 \to \mathfrak{n}/\mathfrak{n}^2 \to 0,$$

and since $[\mathfrak{n}/\mathfrak{n}^2 : k] = \dim A/\mathfrak{p}$, we have $[\mathfrak{p}/\mathfrak{p} \cap \mathfrak{m}^2 : k] = \mathrm{ht}_A \, \mathfrak{p}$.

Thus if $x_1, \ldots, x_p$ are elements of $\mathfrak{p}$ whose images in $\mathfrak{m}/\mathfrak{m}^2$ form a k-basis of $\mathfrak{p}/\mathfrak{p} \cap \mathfrak{m}^2$, then the ideal $(x_1, \ldots, x_p)$ is prime and of height $p = \mathrm{ht}_A \, \mathfrak{p}$; whence $\mathfrak{p} = (x_1, \ldots, x_p)$, qed.

Proposition 23. If $\mathfrak{p}$ is a prime ideal of a regular ring A, then the local ring $A_\mathfrak{p}$ is regular.

Indeed, it follows from the properties proved in part C that $\mathrm{glob \, dim} \, A_\mathfrak{p} \leq \mathrm{glob \, dim} \, A < \infty$.

Proposition 24.   If $\hat{A}$ is the completion of the local ring A for the $\mathfrak{m}$-adic topology, we have the equivalences:

$$A \text{ regular} \iff \hat{A} \text{ regular}.$$

Indeed, $\mathrm{gr}(A) = \mathrm{gr}(\hat{A})$.

This last characterization of regular local rings is very useful, because of the following theorem:

Theorem 10.   If A is a complete local ring, and if A and $k = A/\mathfrak{m}$ have the same characteristic, the following properties are equivalent:
 a) A is regular of dimension n.
 b) A is isomorphic to the formal power series ring $k[[X_1, \ldots , X_n]]$.

The implication b) $\Rightarrow$ a) follows from theorem 9.

Conversely, a) $\Rightarrow$ b): We use the well-known fact that every complete local ring A, with the same characteristic as its residue field k, contains a field k' mapping isomorphically onto k (Cohen, [Co]). For every regular system $\{x_1, \ldots , x_n\}$ of parameters of A, there thus exists a unique homomorphism ϕ from $k'[X_1, \ldots , X_n]$ into A which maps X_i to x_i. Since A is complete, ϕ extends to $k'[[X_1, \ldots , X_n]]$. Since A is regular, the map

$$\mathrm{gr}(\phi) : \mathrm{gr}(k'[[X_1, \ldots , X_n]]) \to \mathrm{gr}(A)$$

is an isomorphism; hence the same is true for ϕ, cf. Chap. II, prop. 6.

Remark.   For a proof of **Cohen's theorem**, see [CaCh], exposé 17, and also [Bour], Chap. IX, §3.

3. Delocalization

It follows from the above that the *regular rings* are the *rings of finite dimension* such that *for every maximal ideal* $\mathfrak{m}$, $A_{\mathfrak{m}}$ *is a regular local ring*, and for these rings the dimension coincide with the global homological dimension:

$$\dim A = \mathrm{glob\,dim}\, A, \qquad \text{if } A \text{ is regular}.$$

Fields and Dedekind domains are the simplest examples of such rings. Apart from these, we have the rings of polynomials according to:

Proposition 25. *If A is a regular ring and $A[X]$ the ring of polynomials in X with coefficients in A, then $A[X]$ is regular and*

$$\operatorname{glob dim} A[X] \;=\; \operatorname{glob dim} A + 1.$$

We first verify the inequality: $\operatorname{glob dim} A[X] \leq \operatorname{glob dim} A + 1$. This is a consequence of:

Lemma 1. *If M is an $A[X]$-module, then*

$$\operatorname{proj dim}_{A[X]}(M) \;\leq\; \operatorname{proj dim}_A M + 1.$$

Let us consider first the case where $M = A[X] \otimes_A N$, N being an A-module (we write $M = N[X]$): since $A[X]$ is A-flat, we have $\operatorname{proj dim}_{A[X]} N[X] \leq \operatorname{proj dim}_A N$, cf. prop. 19. Moreover, it is clear that $\operatorname{proj dim}_A N = \operatorname{proj dim}_A M$.

Now if M is an arbitrary $A[X]$-module, it is also an A-module, and we will let $M[X]$ denote the $A[X]$-module defined by the A-module M.

We have an exact sequence (cf. Bourbaki, *Algèbre VII*, §5.1):

$$0 \to M[X] \xrightarrow{\;\psi\;} M[X] \xrightarrow{\;\phi\;} M \to 0,$$

where

$$\phi\Big(\sum_i X^i \otimes_A m_i \Big) \;=\; \sum_i X^i m_i,$$

and

$$\psi\Big(\sum_i X^i \otimes_A m_i \Big) \;=\; \sum_i X^{i+1} \otimes_A m_i - \sum_i X^i \otimes_A X m_i.$$

Whence

$$
\begin{aligned}
\operatorname{proj dim}_{A[X]} M \;&\leq\; \sup \big(\operatorname{proj dim}_{A[X]} M[X],\, \operatorname{proj dim}_{A[X]} M[X] + 1\big) \\
&=\; \sup \big(\operatorname{proj dim}_A M,\, \operatorname{proj dim}_A M + 1\big) \\
&=\; \operatorname{proj dim}_A M + 1,
\end{aligned}
$$

qed.

Finally, let us show that $\operatorname{glob dim} A[X] \geq \operatorname{glob dim} A + 1$: Indeed, if $\mathfrak{m}$ is an ideal of A such that $\operatorname{ht}_A \mathfrak{m} = \dim A = \operatorname{glob dim} A$, we have

$$
\begin{aligned}
\operatorname{glob dim} A[X] \;=\; \dim A[X] \;&\geq\; \operatorname{ht}_{A[X]}(\mathfrak{m}[X], X) \\
&\geq\; \operatorname{ht}_{A[X]} \mathfrak{m}[X] + 1 \\
&\geq\; \operatorname{ht}_A \mathfrak{m} + 1.
\end{aligned}
$$

Corollary .   If k is a field, $k[X_1, \ldots, X_n]$ is regular.

Since every affine algebra is a quotient of a polynomial ring, we thus recover the properties of chains of prime ideals in affine algebras.

Remark.   Let $A = k[X_1, \ldots, X_n]/\mathfrak{a}$ where $\mathfrak{a}$ is an ideal of the polynomial ring $k[X_1, \ldots, X_n]$. Let X be the corresponding affine variety. One says that X is **non-singular** if A is regular in the sense defined above. When k is perfect, the following are equivalent (cf. e.g. [Bour], Chap. X, §7):
   - X is non-singular;
   - X is smooth over k;
   - $k' \otimes_k A$ is regular for every extension k' of k.

4. A criterion for normality

Theorem 11.   Let A be a noetherian local ring. For A to be normal, it is necessary and sufficient that it satisfies the following two conditions:
 (i) For every prime ideal $\mathfrak{p}$ of A, such that $\mathrm{ht}(\mathfrak{p}) \leq 1$, the local ring $A_\mathfrak{p}$ is regular (i.e. a field or a discrete valuation ring, according to $\mathrm{ht}(\mathfrak{p}) = 0$ or 1).
 (ii) If $\mathrm{ht}(\mathfrak{p}) \geq 2$, we have $\mathrm{depth}(A_\mathfrak{p}) \geq 2$.

Suppose A is normal, and let $\mathfrak{p}$ be a prime ideal of A. If $\mathrm{ht}(\mathfrak{p}) \leq 1$, $A_\mathfrak{p}$ is regular (cf. Chap. III, prop. 9). If $\mathrm{ht}(\mathfrak{p}) \geq 2$, let x be a nonzero element of $\mathfrak{p}A_\mathfrak{p}$; then (*loc. cit.*) every essential prime ideal of $xA_\mathfrak{p}$ in $A_\mathfrak{p}$ is of height 1; thus none of them is equal to $\mathfrak{p}A_\mathfrak{p}$, which shows that $\mathrm{depth}(A_\mathfrak{p}) \geq 2$.

Conversely, suppose that A satisfies (i) and (ii). If we already know that A is a domain, prop. 9 of Chap. III shows that A is normal. In the general case, one first proves that A is *reduced* (i.e. without nonzero nilpotent elements), and then that it is equal to its integral closure in its total ring of fractions. For the details, see [EGA], Chap. IV, th. 5.8.6, or [Bour], Chap. X, p. 21, Rem. 1.

5. Regularity in ring extensions

In this section, A is a noetherian local ring, with residue field k, and B is a noetherian local ring containing A and such that $\mathfrak{m}(A) \subset \mathfrak{m}(B)$. (Note that this condition is satisfied when B is a finitely generated A-module, thanks to Nakayama's lemma.)

Theorem 12.   *If B is regular, and A-flat, then A is regular.*

By prop. 18, we have

$$B \otimes_A \mathrm{Tor}_i^A(k, k) = \mathrm{Tor}_i^B(B \otimes k, B \otimes k) \qquad \text{for all } i.$$

Since B is regular, these modules are 0 for $i > \dim B$. This implies that the same is true for $\mathrm{Tor}_i^A(k, k)$, because of the following lemma:

Lemma 2.   *If M is a finitely generated A-module, then:*

$$B \otimes_A M = 0 \implies M = 0.$$

Indeed, if M were $\neq 0$, it would have a quotient which is isomorphic to $k = A/\mathfrak{m}(A)$ and we would have $B \otimes_A k = 0$, contradicting the fact that $\mathfrak{m}(A)$ is contained in $\mathfrak{m}(B)$.

Hence $\mathrm{Tor}_i^A(k, k) = 0$ for large i, which shows that A is regular.

Theorem 13.   *Assume B is finitely generated as an A-module.*
(a) If A is regular, then: B is A-free $\Leftrightarrow B$ is Cohen-Macaulay.
(b) If B is regular, then: B is A-free $\Leftrightarrow A$ is regular.

Assume A is regular. By prop. 12, B is a Cohen-Macaulay ring if and only if $\mathrm{depth}_A B = \dim A$, i.e. if and only if B is a Cohen-Macaulay A-module. By part ii) of cor. 2 to th. 9, this means that B is A-free. This proves (a) and half of (b). The other half of (b) follows from th. 12.

Appendix I: Minimal Resolutions

In what follows, we let A denote a noetherian local ring, with maximal ideal $\mathfrak{m}$, and residue field k. Every A-module is assumed to be finitely generated. If M is such a module, we write $\overline{M}$ for the k-vector space $M/\mathfrak{m}M$.

1. Definition of minimal resolutions

Let L, M be two A-modules, L being free, and let $u : L \to M$ be a homomorphism. Then u is called **minimal** if it satisfies the following two conditions:

a) u is surjective.

b) $\mathrm{Ker}(u) \subset \mathfrak{m}L$.

It amounts to the same (Nakayama's lemma) to say that $\overline{u} : \overline{L} \to \overline{M}$ is bijective.

If M is given, one constructs a minimal $u : L \to M$ by taking a basis $(\overline{e}_i)$ of the k-vector space $\overline{M} = M/\mathfrak{m}M$, and lifting it to (e_i), with $e_i \in M$.

Now let
$$\ldots \to L_i \xrightarrow{d} \ldots \xrightarrow{d} L_1 \xrightarrow{d} L_0 \xrightarrow{e} M \to 0$$
be a *free resolution* $L_\bullet$ of M. Set:
$$N_i = \mathrm{Im}(L_i \to L_{i-1}) = \mathrm{Ker}(L_{i-1} \to L_{i-2}).$$
Then $L_\bullet$ is called a **minimal free resolution** of M if $L_i \to N_i$ is minimal for every $i \geq 1$, and $e : L_0 \to M$ is minimal as well.

Proposition 1.

(a) *Every A-module M has a minimal free resolution.*

(b) *For a free resolution $L_\bullet$ of M to be minimal, it is necessary and sufficient that the maps $\overline{d} : \overline{L}_i \to \overline{L}_{i-1}$ are zero.*

(a): Choose a minimal homomorphism $e : L_0 \to M$. If N_1 is its kernel, choose a minimal homomorphism $L_1 \to N_1$, etc.

(b): Since $L_\bullet$ is a resolution, the homomorphisms
$$d : L_i \to N_i \quad \text{and} \quad e : L_0 \to M$$
are surjective. For these to be minimal, it is necessary and sufficient that their kernels N_{i+1} (resp. N_1) are contained in $\mathfrak{m}L_i$ (resp. in $\mathfrak{m}L_0$), which means that the boundary operator $\overline{d}$ on $\overline{L}_\bullet$ is zero.

Corollary .   If $L_\bullet = (L_i)$ is a minimal free resolution of M, the rank of L_i is equal to the dimension of the k-vector space $\operatorname{Tor}_i^A(M, k)$.

Indeed, we have:

$$\operatorname{Tor}_i^A(M, k) \cong H_i(L_\bullet \otimes k) = H_i(\overline{L}_\bullet) \cong \overline{L}_i.$$

Remark.   In particular, the rank of L_i is independent of the chosen resolution $L_\bullet$. In fact, it is easy to prove more: any two minimal free resolutions of M are *isomorphic* (non-canonically in general). See e.g. [Eil], or [Eis], §19.1.

2. Application

Let $L_\bullet = (L_i)$ be a minimal free resolution of M, and let $K_\bullet = (K_i)$ be a *free complex*, given with an augmentation $K_0 \to M$. We make the following hypotheses:

(C_0) $\overline{K}_0 \to \overline{M}$ is injective.

(C_i) the boundary operator $d_i : K_i \to K_{i-1}$ maps K_i into $\mathfrak{m}K_{i-1}$ and the corresponding map $\tilde{d}_i : K/\mathfrak{m}K_i \to \mathfrak{m}K_{i-1}/\mathfrak{m}^2 K_{i-1}$ is injective.

Since $L_\bullet$ is a free resolution of M, the identity map $M \to M$ can be extended to a homomorphism of complexes

$$f : K_\bullet \to L_\bullet .$$

Proposition 2.   The map f is injective, and identifies $K_\bullet$ with a direct factor of $L_\bullet$ (as A-modules).

We need to show that the $f_i : K_i \to L_i$ are left-invertible. But, we have the following lemma (whose proof is easy):

Lemma .   Let L and L' be two free A-modules, and let $g : L \to L'$ be a homomorphism. For g to be left-invertible (resp. right-invertible), it is necessary and sufficient that $\overline{g} : \overline{L} \to \overline{L}'$ is injective (resp. surjective).

We thus need to prove that the $\overline{f}_i : \overline{K}_i \to \overline{L}_i$ are injective. We proceed by induction on i:

a) $i = 0$. We use the commutative diagram

$$\begin{array}{ccc} \overline{K}_0 & \longrightarrow & \overline{L}_0 \\ \downarrow & & \downarrow \\ M & \xrightarrow{\ id\ } & M. \end{array}$$

The fact that $\overline{K}_0 \to \overline{M}$ is injective implies that $\overline{K}_0 \to \overline{L}_0$ is injective.

b) $i \geq 1$. We use the commutative diagram

$$\begin{array}{ccc} \overline{K}_i & \longrightarrow & \overline{L}_i \\ \downarrow & & \downarrow \\ \mathfrak{m}K_{i-1}/\mathfrak{m}^2 K_{i-1} & \longrightarrow & \mathfrak{m}L_{i-1}/\mathfrak{m}^2 L_{i-1}. \end{array}$$

Since $f_{i-1} : K_{i-1} \to L_{i-1}$ is left-invertible, so is the map

$$\tilde{f}_{i-1} : \mathfrak{m}K_{i-1}/\mathfrak{m}^2 K_{i-1} \to \mathfrak{m}L_{i-1}/\mathfrak{m}^2 L_{i-1};$$

by the condition (C_i), we conclude that the "diagonal" of the square above is an injective map, whence the injectivity of $\overline{K}_i \to \overline{L}_i$.

Corollary .   The canonical map $H_i(K_{\bullet} \otimes k) \to \mathrm{Tor}_i^A(M, k)$ is injective for every $i \geq 0$.

Indeed, $H_i(K_{\bullet} \otimes k) = H_i(\overline{K}_{\bullet}) = \overline{K}_i$ and $\mathrm{Tor}_i^A(M, k) = \overline{L}_i$; the corollary just reformulates the injectivity of the $\overline{f}_i$.

3. The case of the Koszul complex

From now on, we let $M = k$, the residue field of A.

Proposition 3.   Let $\mathbf{x} = (x_1, \ldots, x_s)$ be a minimal system of generators of $\mathfrak{m}$, and let $K_{\bullet} = K(\mathbf{x}, A)$ be the corresponding Koszul complex. The complex $K_{\bullet}$ (given with the natural augmentation $K_0 \to k$) satisfies the conditions (C_0) and (C_i) of §2.

We have $K_0 = A$ and the map $\overline{A} \to k$ is bijective. The condition (C_0) is thus satisfied. It remains to check (C_i).
Set $L = A^s$; let $(e_1, \ldots, e_s)$ be the canonical basis of L and $(e_1^*, \ldots, e_s^*)$ be the dual basis. We can identify K_i with $\bigwedge^i L$; the boundary map

$$d : \bigwedge^i L \to \bigwedge^{i-1} L$$

is then expressed in the following manner:

$$d(y) = \sum_{j=1}^{j=s} x_j (y \llcorner e_j^*).$$

The symbol $\llcorner$ denotes the *right interior product* (cf. Bourbaki, *Algèbre III*, §11). We need to write $\tilde{d}$ explicitly; for that, we identify $\overline{K}_i$ with $\bigwedge^i \overline{L}$ and $\mathfrak{m} K_{i-1}/\mathfrak{m}^2 K_{i-1}$ with $\mathfrak{m}/\mathfrak{m}^2 \otimes \bigwedge^{i-1} \overline{L}$. Then the formula giving $\tilde{d}$ becomes:

$$\tilde{d}(\overline{y}) = \sum_{j=1}^{j=s} \overline{x}_j (\overline{y} \llcorner \overline{e}_j^*),$$

with obvious notation. Since the $\overline{x}_j$ form a *basis* of $\mathfrak{m}/\mathfrak{m}^2$, the equation $\tilde{d}(\overline{y}) = 0$ is equivalent to $\overline{y} \llcorner \overline{e}_j^* = 0$ for every j, whence $\overline{y} = 0$, qed.

Theorem .   We have $\dim_k \operatorname{Tor}_i^A(k,k) \geq \binom{s}{i}$, with $s = \dim_k \mathfrak{m}/\mathfrak{m}^2$.

Indeed, prop. 3, together with the corollary to prop. 2, shows that the canonical map from $H_i(K_{\bullet} \otimes k) = \overline{K}_i$ to $\operatorname{Tor}_i^A(k,k)$ is injective, and we have $\dim_k \overline{K}_i = \binom{s}{i}$.

Complements.   There are in fact much more precise results (cf. the papers of Assmus [As], Scheja [Sc], Tate [T] quoted in the bibliography):

$\operatorname{Tor}_{\bullet}^A(k,k)$ has a product (the product $\pitchfork$ of [CaE]) which makes it a (skew) commutative, associative k-algebra with a unit element; the squares of elements of odd degree are zero. The isomorphism $\mathfrak{m}/\mathfrak{m}^2 \to \operatorname{Tor}_1^A(k,k)$ extends to a homomorphism of algebras $\phi : \bigwedge \mathfrak{m}/\mathfrak{m}^2 \to \operatorname{Tor}_{\bullet}^A(k,k)$ which is injective [T]; we thus recover the theorem above. The ring A is regular if and only if ϕ is bijective (it is even enough, according to Tate, that the component of ϕ of some degree ≥ 2 is bijective). Moreover, $\operatorname{Tor}_{\bullet}^A(k,k)$ has a co-product [As] which makes it a "Hopf algebra". One can thus apply the structure theorems of Hopf-Borel to it; this gives another proof of the injectivity of ϕ. One also obtains information on the Poincaré series $P_A(T)$ of $\operatorname{Tor}_{\bullet}^A(k,k)$:

$$P_A(T) = \sum_{i=0}^{\infty} a_i T^i, \qquad \text{where } a_i = \dim_k \operatorname{Tor}_i^A(k,k).$$

For example ([T], [As]), A is a "complete intersection" if and only if $P_A(T)$ is of the form $(1+T)^n/(1-T^2)^d$, with $n, d \in \mathbf{N}$; for other analogous results, see [Sc]. In all such cases, $P_A(T)$ turns out to be a rational function of T. In the original version of the Notes, it was asked whether this is always true: "On ignore si $P_A(T)$ est toujours une fonction rationnelle". This question, which some people misread as a *conjecture*, was solved negatively in 1982 by D. J. Anick ([An1,2]), together with the analogous question for loop spaces of simply connected finite complexes.

Appendix II: Positivity of Higher Euler-Poincaré Characteristics

We choose the framework of *abelian categories*. More precisely, let C be an abelian category given with n morphisms $x_1, \ldots, x_n$ from the identity functor into itself. This means that every $E \in C$ is given with endomorphisms $x_1(E), \ldots, x_n(E)$, and that, for every $f \in \mathrm{Hom}_C(E, E')$, we have $x_i(E') \circ f = f \circ x_i(E)$. In particular, the $x_i(E)$ belong to the center of $\mathrm{End}_C(E)$, and they commute with each other.

If $J \subset I = [1, n]$, the subcategory of C which consists of objects E such that $x_i(E) = 0$ for $i \in J$ is written as C_J. We have $C_\emptyset = C$; besides this case, we will have to consider $J = I$ and $J = [2, n]$.

If E is an object in C, the Koszul complex $K(\mathbf{x}, E)$ is defined in an obvious way; its homology groups $H_i(\mathbf{x}, E)$ are objects of C, annihilated by each x_i; in other words, they are elements of C_I.

We now consider the *higher Euler-Poincaré characteristics* formed by means of the $H_i(\mathbf{x}, E)$. First recall how one attaches (according to Grothendieck) a group $K(\mathcal{D})$ to every abelian category $\mathcal{D}$. One first forms the free group $L(\mathcal{D})$ generated by the elements of $\mathcal{D}$; if $0 \to E' \to E \to E'' \to 0$ is an exact sequence in $\mathcal{D}$, one associates to it the element $E - E' - E''$ of $L(\mathcal{D})$; the group $K(\mathcal{D})$ is the quotient of $L(\mathcal{D})$ by the subgroup generated by these elements (for every exact sequence). If $E \in \mathcal{D}$, its image in $K(\mathcal{D})$ is written as $[E]$; the elements of $K(\mathcal{D})$ so obtained are called *positive*; they generate $K(\mathcal{D})$; the sum of two positive elements is a positive element.

This applies to the categories C_J. In particular, let $E \in C$; we have $H_i(\mathbf{x}, E) \in C_I$, and the alternating sum:

$$\chi_i(\mathbf{x}, E) = [H_i(\mathbf{x}, E)] - [H_{i+1}(\mathbf{x}, E)] + \ldots \qquad (i = 0, 1, \ldots)$$

makes sense in the group $K(C_I)$. We can thus ask whether this characteristic χ_i is ≥ 0 (in the sense defined above). We shall see that it is indeed the case if C has the following property:

    (N) *Every $E \in C$ is noetherian, i.e. satisfies the ascending chain condition for subobjects.*

In other words,

Theorem .   If C has property (N), we have $\chi_i(\mathbf{x}, E) \geq 0$ for every $E \in C$ and every $i \geq 0$.

We prove this by induction on n .

a)   *The case $n = 1$*

For simplicity, we write x instead of x_1 . We have

$$H_0(x, E) = \operatorname{Coker} x(E) \quad \text{and} \quad H_1(x, E) = \operatorname{Ker} x(E).$$

The positivity of $\chi_i(x, E)$ is clear for $i > 0$. When $i = 0$, we have to show that

$$\chi_0(x, E) = [\operatorname{Coker} x(E)] - [\operatorname{Ker} x(E)]$$

is a positive element of $K(C_I)$. For $m = 1, 2, \ldots$, let x^m be the m -th power of $x(E)$, and let N_m be the kernel of x^m ; the N_m increase with m . According to (N), the N_m stabilize; let N be their limit, and let $F = E/N$. We have an exact sequence

$$0 \to N \to E \to F \to 0.$$

The additivity of χ_0 implies $\chi_0(x, E) = \chi_0(x, N) + \chi_0(x, F)$. Since $\operatorname{Ker} x(F) = 0$, $\chi_0(x, F)$ is equal to $[H_0(x, F)]$, hence is ≥ 0 . On the other hand, $x(N)$ is nilpotent; this shows that N has a composition series whose successive quotients Q_α are annihilated by x . We have $\chi_0(x, Q_\alpha) = 0$ for every α , hence $\chi_0(x, N) = \sum \chi_0(x, Q_\alpha) = 0$, and we get

$$\chi_0(x, E) = \chi_0(x, F) \geq 0.$$

b)   *Passing from $n - 1$ to n*

According to prop. 1 of part A, we have an exact sequence:

$$0 \to H_0(x_1, H_i(\mathbf{x}', E)) \to H_i(\mathbf{x}, E) \to H_1(x_1, H_{i-1}(\mathbf{x}', F)) \to 0,$$

writing $\mathbf{x}'$ for the sequence $(x_2, \ldots, x_n)$.

Passing to $K(C_I)$, we can thus write:

$$[H_i(\mathbf{x}, E)] = [H_0(x_1, H_i')] + [H_1(x_1, H_{i-1}')],$$

where $H_i' = H_i(\mathbf{x}', E)$. Hence:

$$\begin{aligned}
\chi_i(\mathbf{x}, E) &= [H_1(x_1, H_{i-1}')] \\
&\quad + \sum_{m \geq 0} (-1)^m ([H_0(x_1, H_{i+m}')] - [H_1(x_1, H_{i+m}')]) \\
&= [H_1(x_1, H_{i-1}')] + \sum_{m \geq 0} (-1)^m \chi_0(x_1, H_{i+m}').
\end{aligned}$$

Let $J = [2, n]$. The H_j' belong to C_J . By the induction hypothesis, the element χ_i' of $K(C_J)$ defined by

$$\chi_i' = \sum_{m \geq 0} (-1)^m [H_{i+m}']$$

is ≥ 0, i.e. equal to $[G_i]$ for some object G_i of C_J. Since χ_0 is additive, we have

$$\chi_0(x_1, G_i) = \sum_{m \geq 0} (-1)^m \chi_0(x_1, H'_{i+m}) \quad \text{in } K(C_J).$$

Hence

$$\chi_i(\mathbf{x}, E) = [H_1(x_1, H'_{i-1})] + \chi_0(x_1, G_i).$$

Since $\chi_0(x_1, G_i) \geq 0$ by a), we have $\chi_i(\mathbf{x}, E) \geq 0$, qed.

Example.   Let A be a noetherian local ring, let $x_1, \dots, x_n$ be a system of parameters of A, and let C be the category of finitely generated A-modules (with the endomorphisms defined by the x_i). The category C_I is the category of A-modules annihilated by the x_i; the *length* defines an isomorphism from $K(C_I)$ onto $\mathbf{Z}$, compatible with the order relations. The above theorem thus gives:

Corollary.   If E is a finitely generated A-module and i is ≥ 0, the integer

$$\chi_i(E) = \ell(H_i(\mathbf{x}, E)) - \ell(H_{i+1}(\mathbf{x}, E)) + \dots$$

is ≥ 0.

Remark.   In the case of the above example, one can prove that

$$\chi_i(E) = 0 \quad \text{implies} \quad H_j(\mathbf{x}, E) = 0 \quad \text{for } j \geq i \geq 1.$$

However, the only proof of this fact that I know of is somewhat involved (it consists of reducing to the case where A is a ring of formal power series over a discrete valuation ring or over a field). I do not know if there exists an analogous statement in the framework of abelian categories.

Exercises.
   1)   Assume that C has property (N). Recall that an object of C is *simple* if it is $\neq 0$ and it has no nonzero proper subobject.
   a) Show that every nonzero object of C has a simple quotient.
   b) (Nakayama's lemma) Show the equivalence of the followng three
       properties of C:
       (i) E is simple $\implies x_i(E) = 0$ for $i = 1, \dots, n$.
       (ii) $\mathrm{Coker}(x_i(E)) = 0$ for some $i \implies E = 0$.
       (iii) $H_0(\mathbf{x}, E) = 0 \implies E = 0$.

   2)   Let $x_1, \dots, x_n$ be elements of a commutative noetherian ring A, and let B be an A-algebra. Let C be the category of left B-modules E which are finitely generated as A-modules (the endomorphisms $x_i(E)$ being those given by the x_i). Show that C has property (N), and that it

has properties (i),(ii),(iii) of exerc. 1 if the x_i belong to the radical of A. (Note that this applies in particular when B is the group algebra $A[G]$ of a group G.)

Appendix III: Graded-polynomial Algebras

All the results proved for local rings have analogues for graded algebras over a field. These analogues can be proved directly, or can be deduced from the local statements. We follow the second method.

1. Notation

We consider finitely generated commutative graded algebras over a field k:

$$A = \bigoplus_{n \geq 0} A_n, \qquad \text{with } A_0 = k,$$

together with graded A-modules $M = \bigoplus M_n$ such that $M_{-n} = 0$ for all sufficiently large n.

We put $\mathfrak{m} = \mathfrak{m}(A) = \bigoplus_{n \geq 1} A_n$; it is a maximal ideal of A, with $A/\mathfrak{m} = k$. The completion $\hat{A}_\mathfrak{m}$ of the local ring $A_\mathfrak{m}$ can be identified with the algebra of formal infinite sums:

$$a_0 + a_1 + \ldots + a_n + \ldots, \qquad \text{with } a_n \in A_n \text{ for every } n.$$

One has the following analogue of Nakayama's lemma:

Lemma 1.   If $M = \mathfrak{m}M$, then $M = 0$.

Indeed, if $M \neq 0$, choose n minimal such that $M_n \neq 0$, and note that M_n is not contained in $\mathfrak{m}M$.

Lemma 2.   If $f : M \to M'$ is a homomorphism of graded modules, and if $M/\mathfrak{m}M \to M'/\mathfrak{m}M'$ is surjective, then f is surjective.

This follows from lemma 1, applied to $M'/f(M)$.

A graded A-module M is called **graded-free** if it has a basis made up of homogeneous elements. If one denotes by $A(d)$ a free A-module with basis a homogeneous element of degree d, M is graded-free if and only if it is isomorphic to a direct sum of $A(d_\alpha)$'s.

Lemma 3.   *The following properties are equivalent:*
(i) M *is graded-free.*
(ii) M *if* A*-flat (as a non-graded module).*
(iii) $\operatorname{Tor}_1^A(M, k) = 0$.

It is clear that (i) $\Rightarrow$ (ii) $\Rightarrow$ (iii). To prove (iii) $\Rightarrow$ (i), choose a k-basis $(\tilde{x}_\alpha)$ of homogeneous elements of $M/\mathfrak{m}M$. Put $d_\alpha = \deg(\tilde{x}_\alpha)$ and select a representative x_α of $\tilde{x}_\alpha$ in M_{d_α}. Let L be the direct sum of the free modules $A(d_\alpha)$. The x_α's define a homomorphism $f : L \to M$, which is surjective by lemma 2. Put $N = \operatorname{Ker}(f)$. We have an exact sequence:

$$\operatorname{Tor}_1^A(M, k) \to N/\mathfrak{m}N \to L/\mathfrak{m}L \to M/\mathfrak{m}M \to 0.$$

By construction, $L/\mathfrak{m}L \to M/\mathfrak{m}M$ is an isomorphism, and by (iii) we have $\operatorname{Tor}_1^A(M, k) = 0$. Hence $N/\mathfrak{m}N = 0$, which implies $N = 0$ by lemma 1. Hence M is isomorphic to the graded-free module N, qed.

2. Graded-polynomial algebras (cf. [LIE], Chap. V, §5, and [Bour], Chap. VIII, §6)

We say that A is a **graded-polynomial algebra** (in French: "algèbre graduée de polynômes") if there exist homogeneous elements $P_1, \dots, P_d$ of A, such that the natural map

$$k[X_1, \dots, X_d] \to A,$$

defined by the P_i's, is an isomorphism. If this is the case, the monomials

$$P_1^{\alpha_1} \cdots P_d^{\alpha_d} \qquad \text{with } \sum \alpha_i \deg(P_i) = n$$

make up a basis of A_n. The **Poincaré series** of A,

$$\phi_A(t) = \sum_{n=0}^{\infty} \dim_k(A_n) t^n$$

is equal to

$$\prod_{i=1}^{d} \frac{1}{1 - t^{a_i}}, \qquad \text{with } a_i = \deg(P_i).$$

This shows that the sequence (a_i) is independent of the choice of the P_i's (if numbered so that $a_1 \leq a_2 \leq \dots \leq a_d$). The a_i are called the **basic degrees** of A. Note that A is isomorphic to the symmetric algebra $\operatorname{Sym}(L)$, where L is the graded-free A-module $L = \bigoplus A(a_i)$.

3. A characterization of graded-polynomial algebras

Let $A = \bigoplus_{n \geq 0} A_n$ be a graded algebra satisfying the conditions of §1.

Theorem 1.   *The following properties are equivalent:*
(1) A is a graded-polynomial algebra.
(2) A is isomorphic (as a non-graded algebra) to a polynomial algebra.
(3) A is a regular ring.
(4) The local ring $A_\mathfrak{m}$ is regular.

It is clear that $(1) \Rightarrow (2) \Rightarrow (3) \Rightarrow (4)$. Let us show $(4) \Rightarrow (1)$. First note that $\mathfrak{m}/\mathfrak{m}^2$ is a graded k-vector space of finite dimension. Let $(p_1, \ldots, p_d)$ be a basis of $\mathfrak{m}/\mathfrak{m}^2$ made up of homogeneous elements; let $a_i = \deg(p_i)$ and choose a representative P_i of p_i in A_{a_i}. Let A' be the graded-polynomial algebra $k[X_1, \ldots, X_d]$, with $\deg(X_i) = a_i$, and let $f : A' \to A$ be the unique morphism such that $f(X_i) = P_i$ for every i. One sees easily (by induction on n) that $f(A'_n) = A_n$ for every n, i.e. that f is surjective. The local map $\hat{f} : \hat{A}'_{\mathfrak{m}'} \to \hat{A}_\mathfrak{m}$ (where $\mathfrak{m}' = \mathfrak{m}(A')$) is surjective, and, since $A_\mathfrak{m}$ is regular, it is an isomorphism (use th. 10 of part D §2, for instance [*]). This implies that f is injective. Hence f is an isomorphism, qed.

Remark.   Another property equivalent to $(1), \ldots, (4)$ is:
(5) $\dim(A) = \dim_k(\mathfrak{m}/\mathfrak{m}^2)$.

4. Ring extensions

Let $B = \bigoplus_{n \geq 0} B_n$ be a commutative graded k-algebra containing A, and such that $B_0 = k$.

Theorem 2.   *Assume that B is a graded-polynomial algebra.*
(a) If B is a graded-free A-module, A is a graded-polynomial algebra.
(b) Conversely, if A is a graded-polynomial algebra, and if B is a finitely generated A-module, then B is a graded-free A-module.

In case (a), the local ring $B_{\mathfrak{m}(B)}$ is regular, and is a flat $A_\mathfrak{m}$-module. By th. 12 of part D §5, this implies that $A_\mathfrak{m}$ is regular; hence A is a graded-polynomial algebra, cf. th. 1.

---

[*] Alternative argument: if $\mathrm{Ker}(\hat{f})$ were $\neq 0$, we would have the inequality $\dim \hat{A}_\mathfrak{m} < \dim \hat{A}'_{\mathfrak{m}'}$, but it is clear that both dimensions are equal to $\dim_k(\mathfrak{m}/\mathfrak{m}^2)$.

A similar argument shows that (b) follows from th. 13 of part D §5, combined with lemma 3.

Remark.   In case (b), the A-module B has a basis (b_α) made up of homogeneous elements. Put $e_\alpha = \deg(b_\alpha)$. Since $B = \bigoplus_\alpha Ab_\alpha$, the Poincaré series $\phi_B(t)$ of B is given by

$$(4.1) \qquad \phi_B(t) = (\sum_\alpha t^{e_\alpha})\phi_A(t).$$

If one writes $\phi_A(t)$ and $\phi_B(t)$ as:

$$\phi_A(t) = \prod_{i=1}^d \frac{1}{1-t^{a_i}}, \qquad \phi_B(t) = \prod_{i=1}^d \frac{1}{1-t^{b_i}},$$

with $d = \dim A = \dim B$, one has

$$(4.2) \qquad (\sum_\alpha t^{e_\alpha})\prod_{i=1}^d (1-t^{b_i}) = \prod_{i=1}^d (1-t^{a_i}).$$

Dividing both sides by $(1-t)^d$, and putting $t = 1$ gives:

$$(4.3) \qquad [B : A]\prod b_i = \prod a_i,$$

where $[B : A]$ is the rank of the free A-module B, i.e. the number of elements of the basis (b_α).

Example.   Take $B = k[X_1,\ldots,X_d]$ with the standard grading: $\deg(X_i) = 1$ for $1 \leq i \leq d$. Choose for A the subalgebra of the *symmetric polynomials*. We have $A = k[P_1,\ldots,P_d]$, where $P_1,\ldots,P_d$ are the elementary symmetric polynomials:

$$P_1 = X_1 + \cdots + X_d, \quad \ldots \quad ,P_d = X_1 \cdots X_d.$$

This shows that A is a graded-polynomial algebra with basic degrees $1, 2, \ldots, d$. By th. 2, B is a graded-free A-module. Indeed, one can check that the monomials

$$X_1^{m_1} \cdots X_d^{m_d}, \qquad \text{with } 0 \leq m_i < i,$$

make up an A-basis of B. The identity (4.2) above becomes:

$$(1-t)^d \prod_{i=1}^d (1+t+\cdots+t^{i-1}) = (1-t)(1-t^2)\cdots(1-t^d).$$

5. Application: the Shephard-Todd theorem

We are going to apply th. 2 to the proof of a theorem of Shephard-Todd ([ST]) which is very useful in Lie group theory.

Let V be a k-vector space of finite dimension d, and W a finite subgroup of $GL(V)$ such that
(a) *the order $|W|$ of W is prime to the characteristic of k,*
(b) *W is generated by pseudo-reflections.*

(A **pseudo-reflection** s of V is an automorphism of V such that $\mathrm{Ker}(s-1)$ is a hyperplane of V. Such an automorphism can be written as $x \mapsto x + u(x)v$, where $v \in V - \{0\}$, and u is a non-zero linear form on V.)

Let $B = \mathrm{Sym}\,V$ be the symmetric algebra of V, with the standard grading; it is a graded-polynomial algebra, isomorphic to $k[X_1, \ldots, X_d]$. The group W acts on B. Let $A = B^W$ be the subalgebra of B made up of the elements fixed by W.

Theorem 3 ([ST], see also [Ch3]).   *The algebra A is a graded-polynomial algebra.*

It is easy to see that B is a finitely generated A-module (e.g. because it is integral over A, and generated as an A-algebra by a finite number of elements). By th. 13, it will be enough to prove that it is a graded-free A-module. By lemma 3, this amounts to proving that the k-vector space $E = \mathrm{Tor}_1^A(B,k)$ is 0.

Note that E is a B-module, and that W acts on E.

Lemma 4.
(i) We have $E^W = 0$.
(ii) The group W acts trivially on $E' = E/\mathfrak{m}(B)E$.

Assume Lemma 4. Since $|W|$ is prime to the characteristic of k, the surjective map $E \to E'$ gives a surjection $E^W \to E'^W$. By (i) we have $E^W = 0$. Hence $E'^W = 0$ and by (ii) we have $E' = 0$, hence $E = 0$ by Nakayama's lemma applied to the B-module E.

Proof of part (i) of Lemma 4
Since the order of W is invertible in A, we have

$$\mathrm{Tor}_i^A(M,k)^W = \mathrm{Tor}_i^A(M^W,k)$$

for every A-module M with an A-linear action of W (e.g. use the projector $\frac{1}{|W|}\sum_{w \in W} w$).

By taking $M = B$, and $i = 1$, this gives
$$E^W = \operatorname{Tor}_1^A(B, k)^W = \operatorname{Tor}_1^A(B^W, k) = \operatorname{Tor}_1^A(A, k) = 0.$$

Proof of part (ii) of Lemma 4

Let $s : x \mapsto x + u(x)v$ be a pseudo-reflection belonging to W. The action of s on B is trivial modulo the ideal vB. (Note that v is an element of B_1.) Hence we may factor $s - 1$ on B as:

$$B \xrightarrow{\lambda} B \xrightarrow{\mu} B,$$

where λ is A-linear, and μ is multiplication by v. This gives an analogous factorization of the endomorphism $(s-1)_E$ induced on E by $s - 1$, namely

$$E \xrightarrow{\lambda_E} E \xrightarrow{\mu_E} E,$$

where μ_E is multiplication by v. In particular, the image of $(s-1)_E$ is contained in vE, hence in $\mathfrak{m}(B)E$. This shows that the pseudo-reflections belonging to W act trivially on $E' = E/\mathfrak{m}(B)E$. Since W is generated by pseudo-reflections, this proves (ii).

Remark.   There is a converse to th. 4: if B^W is a *graded-polynomial algebra, then W is generated by pseudo-reflections* (see [ST], and also Bourbaki, [LIE], Chap. V, §5, no. 5).

Exercises   (cf. [LIE], Chap. V).

Let B, A be as in th. 3, and call a_i $(i = 1, \dots, d)$ the basic degrees of A. Assume $\operatorname{char}(k) = 0$.

Let $C = B \otimes_A k = B/\mathfrak{m}(A)B$. It is a graded k-vector space, with an action of W. Let $w \in W$; if $n \geq 0$, denote by $\chi_n(w)$ the trace of w acting on the n-th component of C.

1) Show that $C \otimes_k A$ is isomorphic to B as a graded $A[W]$-module. Deduce that the linear representation of W on C is isomorphic to the regular representation of W.

2) Prove that $|W| = [B : A] = \dim_k C = \prod_{i=1}^d a_i$.

3) Show that $\phi_A(t) \sum_{n \geq 0} \chi_n(w) t^n = \dfrac{1}{\det(1 - tw)}$ for $w \in W$, where

$$\phi_A(t) = \prod_{i=1}^d \frac{1}{1 - t^{a_i}}.$$

4) Show that $\dfrac{1}{|W|} \displaystyle\sum_{w \in W} \chi_n(w) = \begin{cases} 1 & \text{if } n = 0, \\ 0 & \text{if } n > 0. \end{cases}$

5) Use 3) and 4) to prove:

$$\phi_A(t) = \frac{1}{|W|} \sum_{w \in W} \frac{1}{\det(1 - tw)}.$$

6) Show that $\det(t-w)$ divides the polynomial $P_A(t) = \displaystyle\prod_{i=1}^{d} (t^{a_i} - 1)$ for every $w \in W$, and that $P_A(t) = \operatorname*{lcm}_{w \in W} \det(t - w)$.

7) Let V' be a subspace of V and let W' be a finite subgroup of $GL(V')$ generated by pseudo-reflections. Assume that every element of W' is the restriction of some element of W. Let a'_j be the basic degrees of W'.

Show, by using 6), that the polynomial $\displaystyle\prod_j (t^{a'_j} - 1)$ divides the polynomial $\displaystyle\prod_i (t^{a_i} - 1)$. In particular, every a'_j divides some a_i.

Chapter V. Multiplicities

A: Multiplicity of a Module

In this section, A is a commutative noetherian ring; all A-modules are assumed to be finitely generated.

1. The group of cycles of a ring

An element of the free abelian group $Z(A)$ generated by the elements of $\mathrm{Spec}(A)$ is called a **cycle** of A (or of $\mathrm{Spec}(A)$). A cycle Z is **positive** if it is of the form

$$Z = \sum n(\mathfrak{p})\mathfrak{p} \quad \text{with } n(\mathfrak{p}) \geq 0 \text{ for every } \mathfrak{p} \in \mathrm{Spec}(A).$$

Let us assume now that A is local, of dimension n. If $p \geq 0$, let $Z_p(A)$ be the subgroup of $Z(A)$ generated by the prime ideals $\mathfrak{p}$ with $\dim A/\mathfrak{p} = p$. The group $Z(A)$ is the direct sum of its subgroups $Z_p(A)$, for $0 \leq p \leq n$.

The cycles are related to A-modules in the following way: Let $K_p(A)$ be the abelian category of A-modules M such that $\dim_A M \leq p$, $K(A)$ be the category of all A-modules. It is clear that if

$$0 \to M \to N \to P \to 0$$

is an exact sequence of $K(A)$ and if M and P belong to $K_p(A)$, then $N \in K_p(A)$.

Under these conditions, let $M \in K_p(A)$, and let $\mathfrak{q}$ be a prime ideal of A with $\dim A/\mathfrak{q} = p$. Then the module $M_\mathfrak{q}$ over $A_\mathfrak{q}$ is of finite length $\ell(M_\mathfrak{q})$ and this length obviously satisfies the following property: if $0 = M_0 \subset \ldots \subset M_i \subset \ldots \subset M_s = M$ is a composition series of M whose quotients M_i/M_{i-1} are of the form $A/\mathfrak{r}$, where $\mathfrak{r}$ is a prime ideal of A, then it has exactly $\ell(M_\mathfrak{q})$ quotients of the form $A/\mathfrak{q}$.

Thus let $z_p : K_p(A) \to Z_p(A)$ be the function defined by

$$z_p(M) = \sum_{\dim A/\mathfrak{q}=p} \ell(M_\mathfrak{q})\mathfrak{q}.$$

It is clear that z_p is an additive function defined on the category $K_p(A)$ and has values in the ordered group $Z_p(A)$. The function z_p is zero on $K_{p-1}(A)$.

Conversely, it is clear that every additive function on $K_p(A)$ which is zero on $K_{p-1}(A)$ factors through z_p.

If A is a domain, then $Z_n(A) \cong \mathbf{Z}$ for $n = \dim A$, and $z_n(M)$ is the *rank* of the A-module M.

2. Multiplicity of a module

Assume A is local; let $\mathfrak{m}$ be its maximal ideal, and let $\mathfrak{a}$ be an $\mathfrak{m}$-primary ideal. Then, for every nonzero A-module M, the Samuel polynomial $P_\mathfrak{a}(M, X)$ defined in Chap. II is of degree equal to $\dim_A M$. Furthermore, its leading term is of the type $eX^r/r!$, where $r = \dim_A M$ and where e is an integer > 0.

The integer e is, by definition, **the multiplicity** of M for the primary ideal $\mathfrak{a}$. One writes it as $e_\mathfrak{a}(M)$; more generally, p being a positive integer and M a module such that $\dim_A M \leq p$, we set

$$e_\mathfrak{a}(M, p) = \begin{cases} e_\mathfrak{a}(M) & \text{if } \dim_A M = p, \\ 0 & \text{if } \dim_A M < p. \end{cases}$$

It follows from the properties proved in Chap. II that $e_\mathfrak{a}(M, p)$ is an additive function on $K_p(A)$, which is zero on $K_{p-1}(A)$; we thus have the **additivity formula**:

$$e_\mathfrak{a}(M, p) = \sum \ell(M_\mathfrak{q}) e_\mathfrak{a}(A/\mathfrak{q}, p)$$

where $\mathfrak{q}$ runs through the prime ideals with $\dim A/\mathfrak{q} = p$ (or those with $\dim A/\mathfrak{q} \leq p$: it amounts to the same).

In particular, if A is a domain of dimension n, we have

$$e_\mathfrak{a}(M, n) = \text{rank}(M) e_\mathfrak{a}(A).$$

If $\mathfrak{a} = \mathfrak{m}$, $e_\mathfrak{a}(M) = e_\mathfrak{m}(M)$ is called the **multiplicity** of M. In particular $e_\mathfrak{m}(A)$ is the **multiplicity** of the local ring A.

If A is regular, its multiplicity is equal to 1, according to Chap. IV. Conversely, if the multiplicity of A is equal to 1 *and if $\hat{A}$ is a domain*, one can show that A is regular (cf. [Nag3], th. 40.6, and also [Bour], Chap. VIII, p. 108, exerc. 4); an example from Nagata shows that it is not enough to assume that A itself is a domain.

Finally, let $\mathbf{x}$ be an ideal of definition of A, which is generated by $x_1, \ldots, x_n$, where $n = \dim A$. According to theorem 1 of Chap. IV, the

i-th homology group of the Koszul complex $K(\mathbf{x}, M)$ is of finite length $h_i(\mathbf{x}, M)$ for every A-module M and every $i \geq 0$, and we have the formula

$$e_{\mathbf{x}}(M, n) = \sum_{i=0}^{n} (-1)^i h_i(\mathbf{x}, M).$$

B: Intersection Multiplicity of Two Modules

1. Reduction to the diagonal

Let k be an algebraically closed field, let U and V be two algebraic sets of the affine space $\mathbf{A}_n(k) \cong k^n$, and let Δ be the diagonal of the product space $\mathbf{A}_n(k) \times \mathbf{A}_n(k) \cong \mathbf{A}_{2n}(k)$. Then Δ is obviously isomorphic to $\mathbf{A}_n(k)$ and the isomorphism identifies $(U \times V) \cap \Delta$ with $U \cap V$. This reduces the study of the intersection of U and V to the study of the intersection of an algebraic set with a linear variety.

This viewpoint already occurred in prop. 17 of Chap. III (Dimensions of intersections in affine space). In particular, in lemma 7, it helps to view $A \otimes_k A$ as the coordinate ring of $\mathbf{A}_n(k) \times \mathbf{A}_n(k)$, and $A/\mathfrak{p} \otimes_k A/\mathfrak{q}$ and $(A \otimes_k A)/\mathfrak{d}$ as the coordinate rings of $U \times V$ and Δ (U and V irreducible). The isomorphism of $(U \times V) \cap \Delta$ with $U \cap V$ translates into:

$$A/\mathfrak{p} \otimes_A A/\mathfrak{q} \cong (A/\mathfrak{p} \otimes_k A/\mathfrak{q}) \otimes_{(A \otimes_k A)} A, \tag{1}$$

where we identify A with $(A \otimes_k A)/\mathfrak{d}$.

This formula generalizes as follows: let A be a commutative algebra with a unit element over a commutative field k (not necessarily algebraically closed); let M and N be two A-modules, B the k-algebra $A \otimes_k A$, and $\mathfrak{d}$ the ideal of B generated by the $a \otimes 1 - 1 \otimes a$, $a \in A$. Then $(A \otimes_k A)/\mathfrak{d}$ is a k-algebra isomorphic to A, and A is given a B-module structure via this isomorphism. We have the formula ([CaE], Chap. IX, 2.8):

$$\mathrm{Tor}_n^B(M \otimes_k N, A) \cong \mathrm{Tor}_n^A(M, N). \tag{2}$$

So, if

$$\to L_n \xrightarrow{d_n} \ldots \to L_0 \to A \to 0$$

is an $(A \otimes_k A)$-projective resolution of A, the bifunctor

$$(M, N) \mapsto (M \otimes_k N) \otimes_B L_{\bullet}$$

is "resolving", i.e. $\mathrm{Tor}_n^A(M, N)$ is naturally isomorphic to the homology modules of the complex $(M \otimes_k N) \otimes_B L_{\bullet}$. In particular, if A is the polynomial algebra $k[X_1, \ldots, X_n]$ in n variables X_i over k, the Koszul complex $K^B((X_i \otimes 1 - 1 \otimes X_i), B)$ is a free resolution of A, and we have:

$$\mathrm{Tor}_n^A(M, N) \cong H_n(K^B((X_i \otimes 1 - 1 \otimes X_i), M \otimes_k N)). \tag{3}$$

We recover the fact that $k[X_1, \ldots, X_n]$ is regular!

In what follows, the reduction to the diagonal will be used via formula (2) suitably generalized to completed tensor products (see below).

2. Completed tensor products

Let A and B be k-algebras, with k, A, B noetherian, and let $\mathfrak{m}$, $\mathfrak{n}$ denote ideals of A and B respectively such that $A/\mathfrak{m}$ and $B/\mathfrak{n}$ are k-modules of finite length. Let M (resp. N) be a finitely generated A-module (resp. B-module) given with an $\mathfrak{m}$-good filtration (M_p) (resp. $\mathfrak{n}$-good filtration (N_q)). Then, M/M_p and N/N_q are k-modules of finite length for every pair (p, q) of natural numbers.

Then, for every natural number i, the modules $\mathrm{Tor}_i^k(M/M_p, N/N_q)$ form a projective system, and the **completed Tor$_i$** are defined by the formula:

$$\widehat{\mathrm{Tor}}_i^k(M, N) = \varprojlim_{(p,q)} \mathrm{Tor}_i^k(M/M_p, N/N_q). \tag{4}$$

For $i = 0$, we obtain the **completed tensor product** (cf. [Sa3]):

$$M \,\widehat{\otimes}_k\, N = \varprojlim_{(p,q)} (M/M_p \otimes_k N/N_q). \tag{5}$$

The abelian groups thus defined have the following properties:

1) The modules $\widehat{\mathrm{Tor}}_i^k(M, N)$ do not depend on the chosen good filtration for M or N, but only on M and on N (and of course on the maximal ideals of A and B containing $\mathfrak{m}$ and $\mathfrak{n}$).

2) As the diagonal of $\mathbf{N} \times \mathbf{N}$ is a cofinal subset, it suffices to take the projective limit over this diagonal:

$$\widehat{\mathrm{Tor}}_i^k(M, N) \cong \varprojlim_p \mathrm{Tor}_i^k(M/M_p, N/N_p). \tag{6}$$

Similarly we may take the above limit over p and then over q:

$$\widehat{\mathrm{Tor}}_i^k(M, N) \cong \varprojlim_p \varprojlim_q \mathrm{Tor}_i^k(M/M_p, N/N_q)$$

$$\cong \varprojlim_q \varprojlim_p \mathrm{Tor}_i^k(M/M_p, N/N_q).$$

3) The canonical maps from $M \otimes_k N$ into $M/M_p \otimes_k N/N_q$ induce the

maps

$$M \otimes_k N \; \to \; \varprojlim_p (M/M_p) \otimes_k \varprojlim_q (N/N_q) \; \cong \; \hat{M} \otimes_k \hat{N}$$

and $\hat{M} \otimes_k \hat{N} \; \to \; M \,\hat{\otimes}_k\, N,$

and it is clear that $M \,\hat{\otimes}_k\, N$ is naturally isomorphic to the completion of $M \otimes_k N$ for the $(\mathfrak{m} \otimes_k B + A \otimes_k \mathfrak{n})$-adic topology.

4)   The ring $A \,\hat{\otimes}_k\, B$ is complete for the $\mathfrak{r}$-adic topology, where

$$\mathfrak{r} = \mathfrak{m} \,\hat{\otimes}_k\, B + A \,\hat{\otimes}_k\, \mathfrak{n}$$

and the $\widehat{\mathrm{Tor}}_i^k(M, N)$ are complete modules for the $\mathfrak{r}$-adic topology. Since

$$(A \,\hat{\otimes}_k\, B)/\mathfrak{r} = (A/\mathfrak{m}) \otimes_k (B/\mathfrak{n})$$

and $(M \,\hat{\otimes}_k\, N)/\mathfrak{r}(M \,\hat{\otimes}_k\, N) = (M/\mathfrak{m}M) \otimes_k (N/\mathfrak{n}N),$

corollary 3 to proposition 6 of Chap. II is applicable and so $A \,\hat{\otimes}_k\, B$ is noetherian and $M \,\hat{\otimes}_k\, N$ is a finitely generated $(A \,\hat{\otimes}_k\, B)$-module.

Moreover the formula $\frac{1}{1-x} = 1 + x + x^2 + \cdots$ shows that $\mathfrak{r}$ is contained in the radical of $A \,\hat{\otimes}_k\, B$ and the maximal ideals of $A \,\hat{\otimes}_k\, B$ correspond to those of $(A/\mathfrak{m}) \otimes_k (B/\mathfrak{n})$.

5)   If $0 \to M' \to M \to M'' \to 0$ is an exact sequence of finitely generated A-modules, the exact sequences

$$\ldots \to \mathrm{Tor}_n^k(M/\mathfrak{m}^p M, N/\mathfrak{n}^q N)$$
$$\to \mathrm{Tor}_n^k(M''/\mathfrak{m}^p M'', N/\mathfrak{n}^q N)$$
$$\to \mathrm{Tor}_{n-1}^k(M'/M' \cap \mathfrak{m}^p M, N/\mathfrak{n}^q N) \to \ldots$$

give an exact sequence

$$\ldots \to \widehat{\mathrm{Tor}}_n^k(M, N) \to \widehat{\mathrm{Tor}}_n^k(M'', N) \to \widehat{\mathrm{Tor}}_{n-1}^k(M', N) \to \ldots.$$

This follows from the following general property: if

$$\phi : (P_i') \to (P_i) \quad \text{and} \quad \psi : (P_i) \to (P_i'')$$

are two morphisms of projective systems of k-modules over an inductively ordered set, if the P_i' are *artinian*, and if the sequences

$$P_i' \xrightarrow{\phi_i} P_i \xrightarrow{\psi_i} P_i''$$

are exact, then the limit sequence

$$\varprojlim P_i' \to \varprojlim P_i \to \varprojlim P_i''$$

is exact.

[Proof. Let (p_i) be an element of $\mathrm{Ker}(\varprojlim P_i \to \varprojlim P_i'')$. Let E_i be the preimage of p_i in P_i'. The set E_i is non-empty, and has a natural structure of affine k-module. Let $\mathfrak{S}_i$ be the set of all affine submodules of E_i. Using the fact that P_i' is artinian, one checks readily that the family $(\mathfrak{S}_i)$ has properties (i), ... ,(iv) of Bourbaki E.III.58, th. 1, and this implies (*loc. cit.*) that $\varprojlim E_i \neq \emptyset$. Hence (p_i) belongs to the image of $\varprojlim P_i'$ in $\varprojlim P_i$.]

6)   Suppose now that k is a *regular* ring of dimension n and suppose that M, viewed as a k-module, has an M-sequence $\{a_1, \dots, a_r\}$, i.e. that there exist r elements a_i in the radical of k such that a_{i+1} is not a zero-divisor in $M/(a_1, \dots, a_i)M$, $0 \leq i \leq r-1$.

Then $\widehat{\mathrm{Tor}}_i^k(M, N) = 0$ for $i > n - r$. Indeed, it suffices to show this when N is a k-module of finite length, since

$$\widehat{\mathrm{Tor}}_i^k(M, N) = \varprojlim \widehat{\mathrm{Tor}}_i^k(M, N/\mathfrak{q}^n N).$$

The exact sequence

$$0 \to M \xrightarrow{a_1} M \to M/a_1 M \to 0,$$

together with the fact that $\widehat{\mathrm{Tor}}_i^k(M, N) = 0$ if $i > n$, gives the exact sequence:

$$0 \to \widehat{\mathrm{Tor}}_n^k(M, N) \xrightarrow{a_1} \widehat{\mathrm{Tor}}_n^k(M, N).$$

But a power of a_1 annihilates N, hence also $\widehat{\mathrm{Tor}}_n^k(M, N)$. It follows that $\widehat{\mathrm{Tor}}_n^k(M, N) = 0$, which proves our assertion for $r = 1$. If $r > 1$, we have $\widehat{\mathrm{Tor}}_n^k(M, N) = 0$, whence the exact sequence

$$0 \to \widehat{\mathrm{Tor}}_{n-1}^k(M, N) \xrightarrow{a_1} \widehat{\mathrm{Tor}}_{n-1}^k(M, N), \qquad \text{etc} \dots .$$

In the examples we will use, the algebra A has an A-sequence of n elements. It follows that $\widehat{\mathrm{Tor}}_i^k(M', N) = 0$ if M' is A-free and $i > 0$. In this case the functor $M \mapsto \widehat{\mathrm{Tor}}_i^k(M, N)$ is the i-th derived functor of the functor $M \mapsto M \widehat{\otimes}_k N$. This implies that $\widehat{\mathrm{Tor}}_i^k(M, N)$ is a finitely generated $A \widehat{\otimes}_k B$-module.

The machine just constructed will only be used in the following two special cases:

a) k *is a field*, $A \cong B \cong k[[X_1, \dots, X_n]]$:

In this case the $\widehat{\mathrm{Tor}}_i^k$ are zero for $i > 0$. Moreover, $A \widehat{\otimes}_k B$ is isomorphic to the ring of formal power series $C \cong k[[X_1, \dots, X_n; Y_1, \dots, Y_n]]$.

Let now $\mathfrak{q}$ (resp. $\mathfrak{q}'$) be an ideal of definition of A (resp. B), and let $\mathfrak{s}$ be the ideal of C generated by $\mathfrak{q}$ and $\mathfrak{q}'$. Let us put on M,

N, $M \widehat{\otimes}_k N$ the $\mathfrak{q}$-adic, $\mathfrak{q}'$-adic and $\mathfrak{s}$-adic topologies, and consider the corresponding graded modules $\mathrm{gr}(M)$, $\mathrm{gr}(N)$, $\mathrm{gr}(M \widehat{\otimes}_k N)$. The natural map $M \otimes_k N \to M \widehat{\otimes}_k N$ induces a homomorphism

$$\mathrm{gr}(M) \otimes_k \mathrm{gr}(N) \to \mathrm{gr}(M \widehat{\otimes}_k N).$$

One checks easily (cf. [Sa3]) that this homomorphism is bijective. Hence:

$$\dim(M \widehat{\otimes}_k N) = \dim(M) + \dim(N)$$

and

$$e_{\mathfrak{s}}(M \widehat{\otimes} N, \dim M + \dim N) = e_{\mathfrak{q}}(M, \dim M) \cdot e_{\mathfrak{q}'}(N, \dim N).$$

Finally if

$$\ldots \to K_n \to \ldots \to K_0 \to M \to 0$$

and

$$\ldots \to L_n \to \ldots \to L_0 \to M \to 0$$

are A- and B-free resolutions of M and N, $(\sum (K_p \widehat{\otimes}_k L_q)_{p+q=n})_n$ is a C-free resolution of $M \widehat{\otimes}_k N$. In particular if we identify A with the C-module $C/\mathfrak{d}$, where $\mathfrak{d} = (X_1 - Y_1, \ldots, X_n - Y_n)$ we have the equality

$$K_p \otimes_A L_q = (K_p \widehat{\otimes}_k L_q) \otimes_C A,$$

whence the formula of reduction to the diagonal:

$$\mathrm{Tor}_i^A(M, N) \cong \mathrm{Tor}_i^C(M \widehat{\otimes}_k N, A).$$

b) k is a complete discrete valuation ring, $A \cong B \cong k[[X_1, \ldots, X_n]]$:

The letter π denotes a generator of the maximal ideal of k and $\overline{k}$ denotes the residue field $k/\pi k$.

Put:
$$C = A \widehat{\otimes}_k B = k[[X_1, \ldots, X_n; Y_1, \ldots, Y_n]],$$
$$\overline{A} = A/\pi A = \overline{k}[[X_1, \ldots, X_n]],$$
$$\overline{C} = \overline{k}[[X_1, \ldots, X_n; Y_1, \ldots, Y_n]].$$

We have $(M \widehat{\otimes}_k N)/\pi(M \widehat{\otimes}_k N) = (M/\pi M) \widehat{\otimes}_k (N/\pi N)$. It follows that, if π is not a zero-divisor in M and N, we have

$$\dim M \widehat{\otimes}_k N = \dim M + \dim N - 1.$$

Finally, resolving M and N as in a), and taking a C-projective resolution of $A = C/\mathfrak{d}$, we end up with a spectral sequence:

$$\mathrm{Tor}_p^C(A, \widehat{\mathrm{Tor}}_q^k(M, N)) \Longrightarrow \mathrm{Tor}_{p+q}^A(M, N).$$

This spectral sequence degenerates if π is not a zero-divisor in M or N; it gives an isomorphism

$$\mathrm{Tor}_p^C(A, M \widehat{\otimes}_k N) \cong \mathrm{Tor}_p^A(M, N).$$

3. Regular rings of equal characteristic

Let us look at case a) of §2 above, i.e. $A = k[[X_1, \ldots, X_n]]$, where k is a field. Here, the Koszul complex $K^C((X_j - Y_j), C)$ is a free resolution of $A = C/\mathfrak{d}$. If M and N are two finitely generated A-modules, the $\operatorname{Tor}_i^A(M, N)$ may thus be identified with $\operatorname{Tor}_i^C(A, M \widehat{\otimes}_k N)$, i.e. with the homology modules of the Koszul complex $K^C((X_j - Y_j), M \widehat{\otimes}_k N)$:

$$\operatorname{Tor}_i^A(M, N) \cong H_i(K^C((X_j - Y_j), M \widehat{\otimes}_k N)).$$

Theorem 1 of Chap. IV applies to this Koszul complex, and gives the following result:

(*) If $M \otimes_A N$ has finite length, then the $\operatorname{Tor}_i^A(M, N)$ have finite lengths, and the Euler-Poincaré characteristic

$$\chi(M, N) = \sum_{i=0}^{i=n} (-1)^i \ell(\operatorname{Tor}_i^A(M, N))$$

is equal to the multiplicity $e_{\mathfrak{d}}(M \widehat{\otimes}_k N, n)$ of the C-module $M \widehat{\otimes}_k N$ for the ideal $\mathfrak{d}$. Thus

$$\chi(M, N) \geq 0,$$
$$\dim_A M + \dim_A N = \dim_C M \widehat{\otimes}_k N \leq n,$$
$$\chi(M, N) = 0 \text{ if and only if } \dim_A M + \dim_A N < n.$$

This result is easily generalized to the regular rings from algebraic geometry. First of all it is clear that every regular ring A is a direct product of a finite number of regular domains (a noetherian ring A such that, for every prime ideal $\mathfrak{p}$, $A_\mathfrak{p}$ is a domain, is a direct product of a finite number of domains). If A is a domain, then A is called of **equal characteristic** if, for every prime ideal $\mathfrak{p}$, $A/\mathfrak{p}$ and A have the same characteristic. We shall say that a regular ring A is of **equal characteristic** if its "domain components" are of equal characteristic, which is to say if, for every prime ideal $\mathfrak{p}$, the local ring $A_\mathfrak{p}$ is of equal characteristic.

Theorem 1. If A is a regular ring of equal characteristic, M and N are two finitely generated A-modules and $\mathfrak{q}$ a minimal prime ideal of $\operatorname{Supp}(M \otimes_A N)$, then:

(1) $\chi_\mathfrak{q}(M, N) = \sum_{i=0}^{i=\dim A} (-1)^i \ell(\operatorname{Tor}_i^A(M, N)_\mathfrak{q})$ is ≥ 0;
(2) $\dim_{A_\mathfrak{q}} M_\mathfrak{q} + \dim_{A_\mathfrak{q}} N_\mathfrak{q} \leq \operatorname{ht}_A \mathfrak{q}$;
(3) $\dim M_\mathfrak{q} + \dim N_\mathfrak{q} < \operatorname{ht}_A \mathfrak{q}$ if and only if $\chi_\mathfrak{q}(M, N) = 0$.

If we localize at $\mathfrak{q}$, and complete, we have

$$\operatorname{Tor}_i^A(M, N)_\mathfrak{q} = \operatorname{Tor}_i^{A_\mathfrak{q}}(M_\mathfrak{q}, N_\mathfrak{q}) = \operatorname{Tor}_i^{\hat{A}_\mathfrak{q}}(\hat{M}_\mathfrak{q}, \hat{N}_\mathfrak{q}).$$

Hence we may assume that A is complete, and that $\mathfrak{q}$ is its maximal ideal. By Cohen's theorem, A is then isomorphic to $k[[X_1, \ldots , X_n]]$, and we apply $(*)$ above.

Moreover:

Complement. *If* $\mathfrak{a}$ *and* $\mathfrak{b}$ *denote the annihilators of* M *and* N *in* A, $\mathfrak{m}$ *the maximal ideal* $\mathfrak{q}A_\mathfrak{q}$ *of* $A_\mathfrak{q}$, $\mathfrak{c}$ *the ideal generated by* $\mathfrak{a} + \mathfrak{b}$ *in* $A_\mathfrak{q}$, *and if* $\chi_\mathfrak{q}(M, N) > 0$, *we have the inequalities:*

$$e_\mathfrak{m}^{A_\mathfrak{q}}(M_\mathfrak{q}, \dim M_\mathfrak{q}) \cdot e_\mathfrak{m}^{A_\mathfrak{q}}(N_\mathfrak{q}, \dim N_\mathfrak{q}) \leq \chi_\mathfrak{q}(M, N)$$

$$\leq e_\mathfrak{c}^{A_\mathfrak{q}}(M_\mathfrak{q}, \dim M_\mathfrak{q}) \cdot e_\mathfrak{c}^{A_\mathfrak{q}}(N_\mathfrak{q}, \dim N_\mathfrak{q}).$$

Indeed, if k denotes a Cohen subfield of $\hat{A}_\mathfrak{q}$, we have seen that $\chi_\mathfrak{q}(M, N)$ is equal to the multiplicity $e_\mathfrak{d}^C(M \hat{\otimes}_k N, \mathrm{ht}_A \mathfrak{q})$, where

$$C = A_\mathfrak{q} \hat{\otimes}_k A_\mathfrak{q}$$

and where $\mathfrak{d}$ is the ideal of C generated by the $a \hat{\otimes}_k 1 - 1 \hat{\otimes}_k a$, $a \in A_\mathfrak{q}$. But the ideal $\mathfrak{a}_\mathfrak{q} \hat{\otimes}_k A_\mathfrak{q} + A_\mathfrak{q} \hat{\otimes}_k \mathfrak{b}_\mathfrak{q} = \mathfrak{f}$ annihilates $M_\mathfrak{q} \hat{\otimes}_k N_\mathfrak{q}$ and the assertion follows from the inclusions:

$$\mathfrak{m} \hat{\otimes}_k A_\mathfrak{q} + A_\mathfrak{q} \hat{\otimes}_k \mathfrak{m} \supset \mathfrak{d} + \mathfrak{f} \supset \mathfrak{c} \hat{\otimes}_k A_\mathfrak{q} + A_\mathfrak{q} \hat{\otimes}_k \mathfrak{c}.$$

4. Conjectures

It is natural to *conjecture* that theorem 1 extends to *all regular rings*. On this subject, one can make the following remarks:

a) Theorem 1 remains true *without any regularity hypothesis* if M is of the form $A/(x_1, \ldots , x_r)$, the x_i being an A-sequence. Then, indeed, M is of finite homological dimension, the $\mathrm{Tor}_i^A(M, N)$ are the homology modules of the complex $K^A((x_i), N)$, and in particular

$$\chi_\mathfrak{q}(M, N) = e_{\mathbf{x}_\mathfrak{q}}^{A_\mathfrak{q}}(N_\mathfrak{q}, r),$$

where $\mathbf{x}$ is the ideal generated by the x_i.

b) One may assume that A is a *complete* regular local ring. Indeed, we can reduce to this case by localization and completion.

c) One may assume that M and N are of the form $M = A/\mathfrak{p}$, $N = A/\mathfrak{q}$, $\mathfrak{p}$ and $\mathfrak{q}$ being prime ideals of A. Indeed $\chi(M, N)$ is "bi-additive" in M and N, and the general case follows by taking composition series of M and N whose quotients are of the form $A/\mathfrak{p}$, $A/\mathfrak{q}$.

In the case of equal characteristic, we first used b), then a) by reduction to the diagonal. These remarks will allow us to generalize theorem 1:

5. Regular rings of unequal characteristic (unramified case)

Theorem 2.   *Theorem 1 remains true if the hypothesis "A is a regular ring of equal characteristic" is replaced by the more general hypothesis:*
> *A is a regular ring, and for every prime ideal $\mathfrak{p}$ of A,*
> *the local ring $A_\mathfrak{p}$ is of equal characteristic,*
> *or is of unequal characteristic and is unramified.*

(As a matter of fact, it suffices that this property is satisfied when $\mathfrak{p}$ is a *maximal ideal*. Indeed, if A is a regular local ring of unequal characteristic and is unramified, every local ring $A_\mathfrak{p}$ is of the same type or is of equal characteristic.)

Recall that a local ring of unequal characteristic is **unramified** if $p \notin \mathfrak{m}^2$, where p denotes the characteristic of the residue field and $\mathfrak{m}$ the maximal ideal. Cohen has proved (see [Co], [Sa1], and [Bour], Chap. IX,§2) that an unramified regular complete local ring of unequal characteristic is of the form $k[[X_1,\ldots,X_n]]$ where k denotes a complete discrete valuation ring which is unramified (notation from §2, case b). By localization and completion, the proof of theorem 2 is reduced to that of:

Lemma 1.   *If $A = k[[X_1,\ldots,X_n]]$, where k is a complete discrete valuation ring and if M and N are two finitely generated A-modules such that $M \otimes_A N$ is of finite length, then we have:*
(1) $\chi(M,N) = \sum_{i=0}^{n+1}(-1)^i \ell(\operatorname{Tor}_i^A(M,N))$ is ≥ 0;
(2) $\dim M + \dim N \leq \dim A = n+1$;
(3) Moreover $\chi(M,N) \neq 0$ if and only if $\dim M + \dim N = \dim A$.
(Note that we do not assume that k is unramified. The lemma is thus, in a sense, more general than theorem 2.)

In view of remark c) of §4 it suffices to give the proof when M and N are "coprime" (i.e. every endomorphism by scalar multiplication is injective or zero). We thus consider the different cases (π denotes a generator of the maximal ideal of k):

α) π *is not a zero-divisor in M or in N:*
We know (cf. §2,b) that

$$\operatorname{Tor}_i^A(M,N) = \operatorname{Tor}_i^C(A, M \,\widehat{\otimes}_k\, N),$$

where

$$C \cong k[[X_1,\ldots,X_n;Y_1,\ldots,Y_n]]$$

and that

$$\dim_C(M \,\widehat{\otimes}_k\, N) = \dim_A M + \dim_A N - 1.$$

Moreover the Koszul complex $K^C((X_i - Y_i), C)$ is a free resolution of the C-module $A = C/\mathfrak{d}$. Remark a) of §4 applies to $\chi^C(A, M \widehat{\otimes}_k N)$, and proves what we want.

β) π *annihilates* M *and is not a zero-divisor in* N:

Then $\overline{M} = M/\pi M = M$, and M is a module over the ring $\overline{A} = \overline{k}[X_1, \dots, X_n]$. We have a spectral sequence ([CaE], Chap. XVI, §4, 2a and 3a):

$$\mathrm{Tor}_p^{\overline{A}}(M, \mathrm{Tor}_q^A(\overline{A}, N)) \implies \mathrm{Tor}_{p+q}^A(M, N).$$

The exact sequence $0 \to A \xrightarrow{\pi} A \to \overline{A} \to 0$ shows that $\overline{A}$ is of homological dimension 1 over A and that

$$\overline{A} \otimes_A N = N/\pi N,$$

and

$$\mathrm{Tor}_1^A(\overline{A}, N) = {}_\pi N = \mathrm{Ann}_N(\pi)$$
$$= \text{set of elements of } N \text{ annihilated by } \pi.$$

The spectral sequence reduces to the exact sequence:

$$\dots \to \mathrm{Tor}_{i-1}^{\overline{A}}(M, {}_\pi N) \to \mathrm{Tor}_i^A(M, N) \to \mathrm{Tor}_i^{\overline{A}}(M, N/\pi N)$$
$$\to \mathrm{Tor}_{i-2}^{\overline{A}}(M, {}_\pi N) \to \dots,$$

whence $\chi^A(M, N) = \chi^{\overline{A}}(M, N/\pi N) - \chi^{\overline{A}}(M, {}_\pi N)$.

But we assumed that ${}_\pi N = 0$; whence

$$\chi^A(M, N) = \chi^{\overline{A}}(M, N/\pi N) \geq 0,$$
$$\dim_{\overline{A}} M + \dim_{\overline{A}} N/\pi N \leq n$$

and the second inequality is strict if and only if $\chi^{\overline{A}}(M, N/\pi N) = 0$. Since $\dim_A M = \dim_{\overline{A}} M$ and $\dim_{\overline{A}} N/\pi N = \dim_A N/\pi N = \dim_A N - 1$, we get what we want.

γ) π *annihilates* M *and* N:

We again consider M as an $\overline{A}$-module, and N as an A-module; the spectral sequence now gives:

$$\chi^A(M, N) = \chi^{\overline{A}}(M, N/\pi N) - \chi^{\overline{A}}(M, {}_\pi N).$$

But in this case $N/\pi N = {}_\pi N = N$, hence $\chi^A(M, N) = 0$; it remains to check that $\dim_A M + \dim_A N = \dim_{\overline{A}} M + \dim_{\overline{A}} N < n$. But since $M \otimes_A N = M \otimes_{\overline{A}} N$ is an $\overline{A}$-module of finite length, and the lemma has been proved for $\overline{A}$, we have $\dim^{\overline{A}} M + \dim^{\overline{A}} N \leq n$, qed.

6. Arbitrary regular rings

It is not known yet [*] how to extend properties (1) and (3) of th. 1 to such rings. On the other hand, one can prove the inequality (2) (the "dimension formula" of algebraic geometry, cf. Chap. III, prop. 17). More precisely:

Theorem 3.   *Let A be a regular ring, $\mathfrak{p}$ and $\mathfrak{q}$ be two prime ideals of A, $\mathfrak{r}$ a prime ideal of A minimal among those containing $\mathfrak{p} + \mathfrak{q}$. Then:*

$$\mathrm{ht}_A\, \mathfrak{p} + \mathrm{ht}_A\, \mathfrak{q} \geq \mathrm{ht}_A\, \mathfrak{r}.$$

By localizing at $\mathfrak{r}$, one may assume that A is local with maximal ideal $\mathfrak{r}$. In that case, the theorem may be reformulated as follows:
(∗)   *If M and N are finitely generated A-modules, then*

$$\ell(M \otimes_A N) < \infty \implies \dim M + \dim N \leq \dim A.$$

To prove (∗), one may assume that A is complete. By a theorem of Cohen [Co], A can be written as A_1/aA_1 where A_1 is a ring of formal power series over a complete discrete valuation ring, and a is a nonzero element of A_1. If we view M and N as A_1-modules, one shows as in the case γ) of §5 that

$$\chi^{A_1}(M, N) = 0,$$

and lemma 1, applied to A_1, shows that $\dim M + \dim N < \dim A_1$, and

$$\dim M + \dim N \leq \dim A = (\dim A_1) - 1,$$

qed.

Let us also mention the following result:

Theorem 4.   *Let A be a regular local ring of dimension n, let M and N be two nonzero finitely generated A-modules such that $M \otimes_A N$ is of finite length, and let i be the largest integer such that $\mathrm{Tor}_i^A(M, N) \neq 0$. We have:*

$$i = \mathrm{proj\,dim}(M) + \mathrm{proj\,dim}(N) - n. \tag{∗}$$

---

[*] In 1996, O. Gabber has proved property (1), i.e. the fact that $\chi(M, N)$ is ≥ 0. See [Be] and [R4].

Moreover, half of (3) had already been proved in 1985 by P. Roberts (see [R1],[R3]) and H. Gillet–C. Soulé [GiS]; namely

$$\dim M + \dim N < \dim A \implies \chi(M, N) = 0.$$

It is still unknown whether the converse implication is true.

Proof (Grothendieck): Let k be the residue field of A. We are going to determine in two different ways the largest integer r such that the "triple Tor" $\mathrm{Tor}_r^A(M, N, k)$ is $\neq 0$:

a)   The spectral sequence

$$\mathrm{Tor}_p^A(\mathrm{Tor}_q^A(M, N), k) \Longrightarrow \mathrm{Tor}_{p+q}^A(M, N, k)$$

shows that $\mathrm{Tor}_j^A(M, N, k) = 0$ if $j > i + n$, and that

$$\mathrm{Tor}_{i+n}^A(M, N, k) = \mathrm{Tor}_n^A(\mathrm{Tor}_i^A(M, N), k) \neq 0,$$

since $\mathrm{Tor}_i^A(M, N)$ is a nonzero A-module of finite length. Hence $r = n+i$.

b)   The spectral sequence

$$\mathrm{Tor}_p^A(M, \mathrm{Tor}_q^A(N, k)) \Longrightarrow \mathrm{Tor}_{p+q}^A(M, N, k)$$

shows that $r = \mathrm{proj\,dim}(M) + \mathrm{proj\,dim}(N)$ (use the "maximum cycle principle"). Whence $n + i = \mathrm{proj\,dim}(M) + \mathrm{proj\,dim}(N)$, qed.

Corollary .   *The hypotheses being those of theorem 4, in order that*

$$\mathrm{Tor}_i^A(M, N) = 0 \quad \text{for } i > 0,$$

it is necessary and sufficient that M and N are Cohen-Macaulay modules and that $\dim M + \dim N = n$.

We can write the integer i from theorem 4 in the following form:

$$
\begin{aligned}
i &= (\mathrm{proj\,dim}(M) + \dim M - n) + (\mathrm{proj\,dim}(N) + \dim N - n) \\
&\quad + (n - \dim M - \dim N) \\
&= (\dim M - \mathrm{depth}\, M) + (\dim N - \mathrm{depth}\, N) \\
&\quad + (n - \dim M - \dim N).
\end{aligned}
$$

But each term between parentheses is ≥ 0 (for the first two, according to Chap. IV; for the third, according to th. 3). Thus $i = 0$ if and only if each of these terms is zero, whence the result we want.

Remark.   When the hypotheses of the corollary are satisfied, we have

$$\chi(M, N) = \ell(M \otimes_A N);$$

it is probable that the converse is true, and that one has

$$\chi(M, N) < \ell(M \otimes_A N)$$

if either M or N is not Cohen-Macaulay. More generally, one may conjecture that each of the *"higher Euler-Poincaré characteristics"*

$$\chi_r(M, N) = \sum_{i \geq 0} (-1)^i \ell(\mathrm{Tor}_{i+r}^A(M, N)), \qquad r = 1, \dots, n,$$

is ≥ 0, and that $\chi_r = 0$ if and only if each of the $\mathrm{Tor}_{i+r}^A(M, N)$ is zero, cf. Chap. IV, Appendix II. This is at least true in the equal characteristic case, according to Auslander-Buchsbaum. That explains why the Gröbner definition of multiplicities (in terms of $\ell(M \otimes_A N)$) gives the right result only when the varieties are locally Cohen-Macaulay (see the examples constructed by Gröbner himself, [Gröb]).

C: Connection with Algebraic Geometry

1. Tor-formula

Let X be an algebraic variety, defined over a field k. For simplicity, we suppose that k is algebraically closed, and X is irreducible. Let U, V, W be three irreducible subvarieties of X, W being an irreducible component of $U \cap V$. Suppose that W meets the open set of smooth points of X, i.e. that the local ring A of X at W is *regular* (the equivalence "smooth" = "regular" follows from the fact that the ground field k is perfect). Then (cf. part B,§3):

$$\dim U + \dim V \leq \dim X + \dim W. \tag{1}$$

When there is equality in this formula, the intersection is called **proper** at W (and one says that U and V **intersect properly** at W).

Let $\mathfrak{p}_U$ and $\mathfrak{p}_V$ be the prime ideals of the local ring A which correspond to the subvarieties U and V. By hypothesis, A is regular, and $A/(\mathfrak{p}_U + \mathfrak{p}_V)$ has finite length. The Euler-Poincaré characteristic

$$\chi^A(A/\mathfrak{p}_U, A/\mathfrak{p}_V) = \sum_{i=0}^{\dim X} (-1)^i \ell_A(\mathrm{Tor}_i^A(A/\mathfrak{p}_U, A/\mathfrak{p}_V))$$

is defined; this is an integer ≥ 0 (cf. part B).

Theorem 1.
(a) *If U and V do not intersect properly at W, we have $\chi^A(A/\mathfrak{p}_U, A/\mathfrak{p}_V) = 0$.*
(b) *If U and V intersect properly at W, $\chi^A(A/\mathfrak{p}_U, A/\mathfrak{p}_V)$ is > 0 and coincides with the intersection multiplicity $i(X, U \cdot V, W)$ of U and V at W, in the sense of Weil, Chevalley, Samuel (cf. [W],[Ch2],[Sa3]).*

Assertion (a) follows from theorem 1 of part B. We will prove (b) in §4, after having shown that the function $I(X, U \cdot V, W) = \chi^A(A/\mathfrak{p}_U, A/\mathfrak{p}_V)$ satisfies the formal properties of an "intersection multiplicity".

2. Cycles on a non-singular affine variety

Let X be a *non-singular affine* variety, of dimension n, and with coordinate ring A. If $a \in \mathbf{N}$, and if M is an A-module of dimension $\leq a$, the cycle $z_a(M)$ is defined (cf. part A); it is a positive cycle of dimension a, which is zero if and only if $\dim M < a$.

Proposition 1. *Let $a, b, c \in \mathbf{N}$ such that $a + b = n + c$. Let M, N be two A-modules such that:*

$$\dim M \leq a, \qquad \dim N \leq b, \qquad \dim M \otimes_A N \leq c. \qquad (2)$$

Then the cycles $z_a(M)$ and $z_b(N)$ are defined, they intersect properly, and the intersection cycle $z_a(M) \cdot z_b(N)$ (defined by linearity from the function I of §1) coincides with the cycle

$$z_c(\mathrm{Tor}^A(M, N)) = \sum (-1)^i z_c(\mathrm{Tor}_i^A(M, N)). \qquad (3)$$

Let W be an irreducible subvariety of X of dimension c, corresponding to a prime ideal $\mathfrak{r}$ of A; let B be the local ring $A_{\mathfrak{r}}$ (i.e. the local ring of X at W). By definition, the coefficient of W in the cycle $z_c(\mathrm{Tor}^A(M, N))$ is equal to

$$\sum (-1)^i \ell_B(\mathrm{Tor}_i^A(M, N)_{\mathfrak{r}}) = \chi^B(M_{\mathfrak{r}}, N_{\mathfrak{r}}).$$

This coefficient is thus "biadditive" in M and N, and is zero if either $\dim M < a$ or $\dim N < b$, cf. part B, §3. The same is obviously true for the coefficient of W in $z_a(M) \cdot z_b(N)$. We are thus reduced to the case where $M = A/\mathfrak{p}$, $N = A/\mathfrak{q}$, the ideals $\mathfrak{p}$ and $\mathfrak{q}$ being prime and corresponding to irreducible subvarieties U and V of X of respective dimensions a and b. In this case, the coefficient of W in $z_c(\mathrm{Tor}^A(M, N))$ is equal to $\chi^B(B/\mathfrak{p}B, B/\mathfrak{q}B) = I(X, U \cdot V, W)$, qed.

Remarks.

1) Proposition 1 gives a very convenient method for computing the intersection product $z \cdot z'$ of two positive cycles z and z', of dimensions a and b, intersecting properly: choose modules M and N for the cycles z and z' such that $\dim M \otimes N$ has the desired dimension (this is automatically the case if $\mathrm{Supp}(M) = \mathrm{Supp}(z)$ and $\mathrm{Supp}(N) = \mathrm{Supp}(z')$), and the cycle $z \cdot z'$ we want is simply the "cycle of $\mathrm{Tor}(M, N)$", i.e. the alternating sum of the cycles of $\mathrm{Tor}_i(M, N)$.

2) In the case of algebraic varieties which are not necessarily affine, *coherent sheaves* replace modules. If $\mathcal{M}$ is such a sheaf, with $\dim \mathrm{Supp}(\mathcal{M}) \leq a$ (which we also write $\dim \mathcal{M} \leq a$), one defines in an obvious way the *cycle* $z_a(\mathcal{M})$. Proposition 1 remains valid, with

the modules $\mathrm{Tor}_i^A(M, N)$ being replaced by the *sheaves* $\mathfrak{Tor}_i(\mathcal{M}, \mathcal{N})$, the $\mathfrak{Tor}$ being taken over the structural sheaf $\mathcal{O}_X$ of X.

3. Basic formulae

We shall see that the product of cycles, defined by means of the function I from §1 (i.e. by taking the "Tor-formula" as definition), satisfies the fundamental properties of intersection theory; these properties being local, we may suppose that the varieties we consider are affine and non-singular. This will allow us to apply proposition 1 from the preceding section.

a) *Commutativity*
Obvious, because of the commutativity of each Tor_i.

b) *Associativity*
We consider three positive cycles z, z', z'' of respective dimensions a, a', a''. We assume that the products $z \cdot z'$, $(z \cdot z') \cdot z''$, $z' \cdot z''$ and $z \cdot (z' \cdot z'')$ are defined, and we have to prove that

$$(z \cdot z') \cdot z'' = z \cdot (z' \cdot z'').$$

Let A be the coordinate ring of the given ambient variety X, and let n be its dimension. Choose an A-module M with support equal to that of z, and such that $z_a(M) = z$; let M' and M'' be such modules for z' and z''.

The desired formula is proved from the "associativity" of Tor. According to [CaE], p. 347, this associativity is expressed by the existence of the triple Tor $\mathrm{Tor}_i^A(M, M', M'')$ and the two spectral sequences:

$$\mathrm{Tor}_p^A(M, \mathrm{Tor}_q^A(M', M'')) \Longrightarrow \mathrm{Tor}_{p+q}^A(M, M', M'') \tag{4}$$

$$\mathrm{Tor}_p^A(\mathrm{Tor}_q^A(M, M'), M'') \Longrightarrow \mathrm{Tor}_{p+q}^A(M, M', M''). \tag{5}$$

Set $c = a + a' + a'' - 2n$, and $b = a' + a'' - n$. Since the intersections considered are proper, we have

$$\dim M' \otimes M'' \leq b \quad \text{and} \quad \dim M \otimes M' \otimes M'' \leq c.$$

One can thus define the cycles

$$y_q = z_b(\mathrm{Tor}_q^A(M', M'')),$$

$$x_{p,q} = z_c(\mathrm{Tor}_p^A(M, \mathrm{Tor}_q^A(M', M''))),$$

$$x_i = z_c(\mathrm{Tor}_i^A(M, M', M'')).$$

The invariance of Euler-Poincaré characteristics through a spectral sequence, applied to (4), gives:

$$\sum_i (-1)^i x_i = \sum_{p,q} (-1)^{p+q} x_{p,q}.$$

But proposition 1 shows that

$$\sum_p (-1)^p x_{p,q} = z \cdot y_q \quad \text{and} \quad \sum_q (-1)^q y_q = z' \cdot z''.$$

Thus

$$\sum_i (-1)^i x_i = z \cdot (z' \cdot z'').$$

The same argument, applied to (5), gives

$$\sum_i (-1)^i x_i = (z \cdot z') \cdot z'',$$

whence the associativity formula we want.

c) *Product formula*

Consider two non-singular varieties X and X', and two cycles z_1, z_2 (resp. z_1', z_2') on X (resp. X'). We suppose that $z_1 \cdot z_2$ and $z_1' \cdot z_2'$ are defined. Then the product cycles $z_1 \times z_1'$ and $z_2 \times z_2'$ (on $X \times X'$) intersect properly, and we have:

$$(z_1 \times z_1') \cdot (z_2 \times z_2') = (z_1 \cdot z_2) \times (z_1' \cdot z_2'). \tag{6}$$

Indeed, we may assume that these cycles are ≥ 0, and that X and X' are affine, with coordinate rings A and A'. If M_1, M_2, M_1', M_2' are modules corresponding to z_1, z_2, z_1', z_2', one checks that the cycle associated to $M_1 \otimes_k M_1'$ (viewed as a module over the ring $B = A \otimes_k A'$ of $X \times X'$) is equal to $z_1 \times z_1'$ [this fact could even be taken as the definition of the direct product of cycles]. The formula to be proved then follows from the "Künneth formula":

$$\operatorname{Tor}_h^B(M_1 \otimes M_1', M_2 \otimes M_2') = \bigoplus_{i+j=h} \operatorname{Tor}_i^A(M_1, M_2) \otimes_k \operatorname{Tor}_j^{A'}(M_1', M_2').$$

d) *Reduction to the diagonal*

Let Δ be the diagonal of $X \times X$. We have to show the formula

$$z_1 \cdot z_2 = (z_1 \times z_2) \cdot \Delta, \tag{7}$$

valid when the cycles z_1 and z_2 intersect properly.

Let A be the coordinate ring of X, and let $B = A \otimes_k A$ be that of $X \times X$. If M_1 and M_2 are modules corresponding to z_1 and z_2 respectively, we have

$$\operatorname{Tor}_i^A(M_1, M_2) = \operatorname{Tor}_i^B(M_1 \otimes_k M_2, A),$$

cf. part B, §1. Formula (7) then follows by taking the alternating sum of the cycles on both sides.

4. Proof of theorem 1

We want to show that the functions I and i coincide. We begin by treating the case where U is a *complete intersection in* W; this means that the ideal $\mathfrak{p}_U$ of §1 is generated by h elements $x_1, \ldots, x_h$, with $h = \dim X - \dim U = \dim V - \dim W$.

We have $\operatorname{Tor}_i^A(A/\mathfrak{p}_U, A/\mathfrak{p}_V) = H_i(\mathbf{x}, A/\mathfrak{p}_V)$, cf. Chap. IV, cor. 2 to prop. 3. Hence

$$\chi^A(A/\mathfrak{p}_U, A/\mathfrak{p}_V) = \sum_i (-1)^i \ell(H_i(\mathbf{x}, A/\mathfrak{p}_V)),$$

and by theorem 1 of Chap. IV, this gives

$$\chi^A(A/\mathfrak{p}_U, A/\mathfrak{p}_V) = e_{\mathbf{x}}(A/\mathfrak{p}_V),$$

where $\mathbf{x}$ denotes the ideal of $A/\mathfrak{p}_V$ generated by the images of the x_i. But according to [Sa4], p. 83, the multiplicity $e_{\mathbf{x}}(A/\mathfrak{p}_V)$ is equal to $i(X, U \cdot V, W)$, which proves the equality $I = i$ in this case.

The general case reduces to the previous one, by using *reduction to the diagonal*, which is valid for both I and i. Since Δ is non-singular, it is locally a complete intersection, and the hypotheses of the preceding case are satisfied, qed.

5. Rationality of intersections

For simplicity we restrict ourselves to the case where X is an affine variety with coordinate ring A over k. Let k_0 be a subfield of k. We say (in Weil style) that X is *defined over* k_0 if one has chosen a k_0-subalgebra A_0 of A such that $A = A_0 \otimes_{k_0} k$.

Let M_0 be an A_0-module (finitely generated, as usual), with $\dim M_0 \leq a$. We can view $M_0 \otimes_{k_0} k$ as an A-module, and we have $\dim(M_0 \otimes k) \leq a$, which allows us to define the cycle $z_a(M_0 \otimes k)$. A cycle z of dimension a on X is called **rational over** k_0 if it is the difference of two cycles $z_a(M_0 \otimes k)$ and $z_a(M_0' \otimes k)$ obtained in this way. The abelian group of cycles rational over k_0 has a basis consisting of the "prime cycles" $z_a(A_0/\mathfrak{p}_0 \otimes k)$, where $\mathfrak{p}_0$ ranges over the set of prime ideals of A_0 such that $\dim(A_0/\mathfrak{p}_0) = a$. This definition of the rationality of cycles is equivalent to that given by Weil in [W]; this non-trivial fact can be proved by interpreting the "order of inseparability" which appears in Weil's book in terms of tensor products of fields (cf. [ZS], Chap. III, p. 118, th. 38).

Theorem 2 (Weil). *Let z and z' be two cycles on X, rational over k_0, and such that $z \cdot z'$ is defined. Then $z \cdot z'$ is rational over k_0.*

We can assume that z and z' are positive, thus corresponding to A_0-modules M_0 and M_0'. The theorem then follows from the formula:

$$\operatorname{Tor}_i^A(M_0 \otimes k, M_0' \otimes k) = \operatorname{Tor}_i^{A_0}(M_0, M_0') \otimes k.$$

6. Direct images

Let $f : X \to Y$ be a morphism of algebraic varieties (over an algebraically closed field k, to fix ideas), and let z be a cycle on X of dimension a. The **direct image** $f_*(z)$ of z is defined by linearity, starting from the case where $z = W$, an irreducible subvariety of X. In this case, set:

$$f_*(W) = \begin{cases} 0 & \text{if the closure } W' \text{ of } f(W) \text{ has dimension } < a, \\ dW' & \text{if } \dim W' = a, \text{ where } d = [k(W) : k(W')] \text{ is} \\ & \text{the degree of the map } f : W \to W'. \end{cases}$$

This operation preserves dimension. It is mainly interesting when f is *proper* (do not confuse the properness of a morphism with that of an intersection!), because of the following result:

Proposition 2. Let $f : X \to Y$ be a proper morphism, let z be a cycle on X of dimension a, and let $\mathcal{M}$ be a coherent sheaf on X such that $z_a(\mathcal{M}) = z$. Let $R^q f(\mathcal{M})$ be the q-th direct image of $\mathcal{M}$, which is a coherent sheaf on Y ([EGA], Chap. III, th. 4.1.5).
(a) We have $\dim R^0 f(\mathcal{M}) \le a$ and $\dim R^q f(\mathcal{M}) < a$ for $q \ge 1$.
(b) We have

$$f_*(z) = z_a(R^0 f(\mathcal{M})) = \sum_q (-1)^q z_a(R^q f(\mathcal{M})).$$

The proof is done by reducing to the case where the restriction of f to the support of z is a *finite* morphism, in which case the $R^q f(\mathcal{M})$ are zero for $q \ge 1$.

7. Pull-backs

They can be defined in diverse situations. We consider only the following:
Let $f : X \to Y$ be a morphism, with Y being *non-singular*, and let x and y be cycles on X and Y respectively. Set $|x| = \operatorname{Supp}(x)$ and $|y| = \operatorname{Supp}(y)$. Then:

$$\dim |x| \cap f^{-1}(|y|) \ge \dim |x| - \operatorname{codim} |y|.$$

The "proper" case is that where there is equality. In that case, one defines an **intersection cycle** $x \cdot_f y$ with support contained in $|x| \cap f^{-1}(|y|)$ by either one of the following methods:

a) *Reduction to a standard intersection*: assume X to be affine (the problem being local), which allows it to be embedded in a non-singular variety V, for example an affine space. The map $z \mapsto (z, f(z))$ embeds X into $V \times Y$, and thus allows us to identify any cycle x on X with a cycle $\gamma(x)$ on $V \times Y$. Then one defines $x \cdot_f y$ as the unique cycle on X such that:

$$\gamma(x \cdot_f y) = \gamma(x) \cdot (V \times y), \tag{8}$$

the intersection product of the right hand side being computed on the non-singular variety $V \times Y$. One checks that the result obtained is independent of the chosen embedding $X \to V$.

b) Choose *coherent sheaves* $\mathcal{M}$ and $\mathcal{N}$ over X and Y of respective cycles x and y, and define $x \cdot_f y$ as the alternating sum of the cycles of the sheaves $\mathfrak{Tor}_i(\mathcal{M}, f^*(\mathcal{N}))$, the $\mathfrak{Tor}_i$ being taken over $\mathcal{O}_X$ (and being sheaves on X); since Y is non-singular, the $\mathfrak{Tor}_i$ are zero for $i > \dim Y$, and the alternating sum is finite.

Special case: Take $x = X$. The cycle $x \cdot_f y$ is then written $f^*(y)$ and is called the **pull-back** of y. Recall the conditions under which it is defined:
  i) Y is non-singular,
  ii) $\operatorname{codim} f^{-1}(|y|) = \operatorname{codim} |y|$.
  No hypothesis on X is necessary.

Remarks.
  1) When X is non-singular, we have

$$x \cdot_f y = x \cdot f^*(y), \tag{9}$$

provided that both sides are defined.
  2) The special case when Y is a *line* is the starting point of the theory of linear equivalence of cycles.

Projection formula
  It is the formula:

$$f_*(x \cdot_f y) = f_*(x) \cdot y, \tag{10}$$

valid when f is proper and both sides are defined.

The proof can be done by introducing sheaves $\mathcal{M}$ and $\mathcal{N}$ corresponding to the cycles x and y, and using two spectral sequences with the same ending and with the E_2 terms being respectively

$$R^q f(\mathfrak{Tor}_p(\mathcal{M}, \mathcal{N})) \quad \text{and} \quad \mathfrak{Tor}_i(R^j f(\mathcal{M}), \mathcal{N}),$$

the $\mathfrak{Tor}$ being taken over $\mathcal{O}_Y$ (cf. [EGA], Chap. III, prop. 6.9.8).
  When X is non-singular, this formula takes the standard form:

$$f_*(x \cdot f^*(y)) = f_*(x) \cdot y. \tag{11}$$

Exercises   (see also [F], Chap. 8).

1) Let $Z \xrightarrow{g} Y \xrightarrow{f} X$, and let x, y, z be positive cycles on X, Y, Z. Suppose that X and Y are non-singular. Prove the following formula (valid whenever the products which appear are defined):

$$z \cdot_g (y \cdot_f x) = (z \cdot_g y) \cdot_{fg} x = (z \cdot_{fg} x) \cdot_g y. \tag{12}$$

Recover (for $f = g = 1$) the associativity and commutativity of the intersection product. For $X = Y$, $f = 1$, deduce the formula:

$$g^*(x \cdot y) = g^*(x) \cdot_g y, \tag{13}$$

whence $g^*(x \cdot y) = g^*(x) \cdot g^*(y)$ when Z is non-singular.

2)   Same hypotheses as in 1), with the difference that Y may be singular, but that g is *proper* (it suffices that its restriction to $\mathrm{Supp}(z)$ is so). Prove the formula:

$$g_*(z \cdot_{fg} x) = g_*(z) \cdot_f x, \tag{14}$$

valid when both sides are defined. (For $f = 1$, one recovers (10).)

3)   Give the conditions of validity for the formula:

$$(y_1 \times y_2) \cdot_{f_1 \times f_2} (x_1 \times x_2) = (y_1 \cdot_{f_1} x_1) \times (y_2 \cdot_{f_2} x_2). \tag{15}$$

4)   Let $f : Y \to X$, $f' : Y \to X'$, with X, X' non-singular, let $g = (f, f') : Y \to X \times X'$. Let x, x', y be cycles on X, X', Y. Give conditions of validity for the formula:

$$(y \cdot_f x) \cdot_{f'} x' = (y \cdot_{f'} x') \cdot_f x = y \cdot_g (x \times x').$$

5)   Let $f : Y \to X$ and $g : Z \to X$, with X non-singular. Let y, z be cycles on Y, Z. Define (under the usual properness conditions) a "fiber product" $y \cdot_X z$, which is a cycle on the fiber product $Y \times_X Z$ of Y and Z over X. What does this give when $g = 1$? And when X is reduced to a point?

8. Extensions of intersection theory

It is clear that the "Tor-formula" allows us to define the intersection of two cycles in more general cases than those of classical algebraic geometry. For example:

i) It applies to *analytic* (or *formal*) *spaces*. There is no difficulty, since every local ring which is involved is of equal characteristic. In the case of *complex* analytic spaces, the intersection product so obtained coincides with that defined topologically by Borel-Haefliger, cf. [BoH]; this is proved by reduction to the "elementary" case 4.10 of their paper.

ii) It applies to every *regular scheme* X provided that the conjectures of part B have been checked for the local rings of such schemes; this is especially the case when these local rings are *of equal characteristic*. [Even when X is a scheme of finite type over a field k, this gives an intersection theory a little more general than the usual one; indeed, if k is not perfect, it may happen that X is regular without being *smooth* over k; but Weil's theory applies only to the smooth case.]

iii) More generally, intersection theory applies to every scheme X which is *smooth over a discrete valuation ring* C. One can indeed show that the local rings of X satisfy the conjectures from part B [the proof is done by a process of reduction to the diagonal which is analogous to — and simpler than — the one used in part B, §5]. This case is important, because it gives the *reduction of cycles* of Shimura, [Sh]. We briefly indicate how:

Let k (resp. K) be the residue field of C (resp. its field of fractions). The scheme X is a disjoint sum of its closed subscheme $X_k = X \otimes_C k$ and its open subscheme $X_K = X \otimes_C K$; the scheme X_k is of finite type over k (it is an "algebraic variety" over k); similarly, X_K is of finite type over K. One sometimes says, rather incorrectly, that X_k is the **reduction** of X_K.

Every cycle on X_k defines, by injection, a cycle on X of the same dimension; every cycle z of dimension a on X_K defines by closure a cycle $\bar{z}$ of dimension $a+1$ on X. The group $Z_n(X)$ of cycles on X of dimension n thus decomposes into a direct sum:

$$Z_n(X) \;=\; Z_n(X_k) \oplus Z_{n-1}(X_K).$$

The projection $Z_n(X) \to Z_{n-1}(X_K)$ is given by the *restriction* of cycles. From the point of view of sheaves, the cycles from $Z(X_k)$ correspond to coherent sheaves $\mathcal{M}$ over X which are annihilated by the uniformizer π of C; those from $Z(X_K)$ correspond to coherent sheaves $\mathcal{M}$ which are flat over C (i.e. torsion-free); this decomposition into two types has already played a role in part B, §5.

Now let $z \in Z_n(X_K)$, and let $\bar{z}$ be its closure. We can view X_k as a cycle of codimension 1 on X. The intersection product

$$\tilde{z} \;=\; X_k \cdot \bar{z} \qquad \text{(computed on } X\text{)}$$

belongs to $Z_n(X_k)$; it is called the **reduction of the cycle** z. Moreover this operation can be defined without speaking of intersections (and without any hypothesis of smoothness or regularity); from the point of view of sheaves, it amounts to associating to every coherent sheaf $\mathcal{M}$ that is flat over C the sheaf $\mathcal{M}/\pi\mathcal{M}$. The hypothesis of smoothness comes only for proving the formal properties of the operation of reduction: compatibility with products, direct images, intersection products; the proofs can be done, as in the preceding sections, by working with identities between sheaves, or, at worst, with spectral sequences.

The *intersection theory on* X gives more than the mere *reduction of cycles*. Thus if x and x' are cycles on X_K, the component of $\bar{x} \cdot \bar{x}'$

in $Z(X_k)$ gives an interesting invariant of the pair x, x' (assuming, of course, that the intersection of $\bar{x}$ and $\bar{x}'$ is proper); this invariant is related to the "local symbols" introduced by Néron, [Né].

Bibliography

[AM] M. F. Atiyah and I. G. Macdonald, *Introduction to Commutative Algebra*, Addison-Wesley, 1969.

[An1] D. J. Anick, *A counterexample to a conjecture (sic) of Serre*, Ann. of Math., **115**, 1982, 1–33.

[An2] D. J. Anick, *A note on the paper "A counterexample to a conjecture of Serre"*, Ann. of Math., **116**, 1982, 661.

[As] E. Assmus, Jr, *On the homology of local rings*, Illinois J. Math., **3**, 1959, 187–199.

[Au1] M. Auslander, *Modules over unramified regular local rings*, Proc. Int. Congress Stockholm (1962), 230–233.

[Au2] ———, *On the purity of the branch locus*, Amer. J. Math., **84**, 1962, 116–125.

[AuB1] M. Auslander and D. Buchsbaum, *Homological dimension in local rings*, Trans. Amer. Math. Soc., **85**, 1957, 390–405.

[AuB2] ———, *Codimension and multiplicity*, Ann. of Math., **68**, 1958, 625–657 (Errata, **70**, 1959, 395–397).

[AuB3] ———, *Unique factorization in regular local rings*, Proc. Nat. Acad. Sci. USA., **45**, 1959, 733–734.

[Ba] H. Bass, *On the ubiquity of Gorenstein rings*, Math. Zeit., **82**, 1963, 8–28.

[Be] P. Berthelot, *Altérations de variétés algébriques (d'après A. J. de Jong)*, Séminaire Bourbaki 1995–1996, no. 815 (= Astérisque, **241**, 273–311).

[Bo] A. Borel, *Sur la cohomologie des espaces fibrés principaux et des espaces homogènes de groupes de Lie compacts*, Ann. of Math., **57**, 1953, 116–207.

[BoH] A. Borel et A. Haefliger, *La classe d'homologie fondamentale d'un espace analytique*, Bull. Soc. Math. France, **89**, 1961, 461–513.

[BoSe] A. Borel et J-P. Serre, *Le théorème de Riemann-Roch (d'après A. Grothendieck)*, Bull. Soc. Math. France, **86**, 1958, 97–136.

[Bour] N. Bourbaki, *Algèbre Commutative*, Hermann-Masson, Paris, 1961–1998.

124   Bibliography

[BrH]   W. Bruns and H. J. Herzog, *Cohen-Macaulay Rings*, 2nd ed., Cambridge Univ. Press, Cambridge, 1998.

[CaCh]   H. Cartan et C. Chevalley, *Géométrie algébrique*, Séminaire ENS, 1956.

[CaE]   H. Cartan and S. Eilenberg, *Homological Algebra*, Princeton Math. Ser. **19**, Princeton, 1956.

[Ch1]   C. Chevalley, *On the theory of local rings*, Ann. of Math., **44**, 1943, 690–708.

[Ch2]   ———, *Intersections of algebraic and algebroid varieties*, Trans. Amer. Math. Soc., **57**, 1945, 1–85.

[Ch3]   ———, *Invariants of finite groups generated by reflections*, Amer. J. Math., **77**, 1955, 778–782.

[Co]   I. Cohen, *On the structure and ideal theory of complete local rings*, Trans. Amer. Math. Soc., **59**, 1946, 54–106.

[D]   P. Dubreil, *La fonction caractéristique de Hilbert*, Colloque d'Alg. et Th. des Nombres, CNRS, Paris, 1950, 109–114.

[Eil]   S. Eilenberg, *Homological dimension and syzygies*, Ann. of Math:, **64**, 1956, 328–336.

[Eis]   D. Eisenbud, *Commutative Algebra with a view toward Algebraic Geometry*, GTM 150, Springer-Verlag, 1994.

[F]   W. Fulton, *Intersection Theory*, Springer-Verlag, 1984.

[Gab]   P. Gabriel, *Des catégories abéliennes*, Bull. Soc. Math. France, **90**, 1962, 323–448.

[Gae]   F. Gaeta, *Quelques progrès récents dans la classification des variétés algébriques d'un espace projectif*, Deuxième Colloque de Géom. Alg., Liège, 1952, 145–183.

[GiS]   H. Gillet and C. Soulé, *Intersection theory using Adams operations*, Invent. math., **90**, 1987, 243-277 (see also C.R.A.S., **300**, 1985, 71–74).

[Gröb]   W. Gröbner, *Moderne algebraische Geometrie*, Springer-Verlag, 1949.

[Groth1]   A. Grothendieck, *Sur quelques points d'algèbre homologique*, Tôhoku Math. Journ., **9**, 1957, 119–221.

[Groth2]   ———, *Sur quelques propriétés fondamentales en théorie des intersections*, Séminaire C. Chevalley, *Anneaux de Chow et applications*, Paris, 1958.

[EGA]   ———, *Eléments de géométrie algébrique (rédigés avec la collaboration de J. Dieudonné)*, Publ. Math. IHES, **4, 8, 11, 17, 20, 24, 28, 32**, 1960–1967.

[SGA 2]   ———, *Séminaire de géométrie algébrique (notes prises par un groupe d'auditeurs)*, Exposés I à XIII, IHES, Paris, 1962; (SGA 2, Masson-North-Holland, 1968).

[H1]   M. Hochster, *Topics in the homological theory of modules over commutative rings*, Regional Conf. Series in Math., **24**, 1975.

[H2]   ———, *Cohen-Macaulay rings and modules*, Proc.   Int.
            Congress Helsinki (1978), vol. 1, 291–298.

[J]   P. Jaffard, *Théorie de la dimension dans les anneaux de
            polynômes*, Mém. Sci. Math., **146**, Gauthier-Villars, Paris,
            1960.

[Ko]   J.-L. Koszul, *Sur un type d'algèbres différentielles en rapport
            avec la transgression*, Colloque de Topologie, Bruxelles, 1950,
            73–81.

[Kr1]   W. Krull, *Idealtheorie*, Springer-Verlag, 1935.

[Kr2]   ———, *Dimensionstheorie in Stellenringen*, J. reine ang.
            Math., **179**, 1938, 204–226.

[Kr3]   ———, *Zur Theorie der kommutativen Integritätsbereiche*, J.
            reine ang. Math., **192**, 1954, 230–252.

[Le1]   C. Lech, *Note on multiplicities of ideals*, Ark. för Math., **4**,
            1959, 63–86.

[Le2]   ———, *Inequalities related to certain couples of local rings*,
            Acta Math., **112**, 1964, 69–89.

[Li]   S. Lichtenbaum, *On the vanishing of* Tor *in regular local rings*,
            Illinois J. Math., **10**, 1966, 220–226.

[LIE]   N. Bourbaki, *Groupes et Algèbres de Lie*, Chap. I–IX,
            Hermann-Masson, Paris, 1961–1982.

[Mac]   F. Macaulay, *Algebraic Theory of Modular Systems*, Cambridge
            Tract **19**, Cambridge, 1916; second edition (with a new intro-
            duction by P. Roberts), Cambridge, 1994.

[Mat]   H. Matsumura, *Commutative Algebra*, Benjamin, New York,
            1970.

[Nag1]   M. Nagata, *Note on integral closure of Noetherian domains*,
            Mem. Univ. Kyoto, **28**, 1953, 121–124.

[Nag2]   ———, *The theory of multiplicity in general local rings*, Symp.
            Tokyo-Nikko, 1955, 191–226.

[Nag3]   ———, *Local Rings*, Interscience Publ., New York, 1962.

[Nas1]   H-J. Nastold, *Über die Assoziativformel und die Lechsche
            Formel in der Multiplizitätstheorie*, Archiv der Math., **12**, 1961,
            105–112.

[Nas2]   ———, *Zur Serreschen Multiplizitätstheorie in der arithmetis-
            chen Geometrie*, Math. Ann., **143**, 1961, 333–343.

[Né]   A. Néron, *Quasi-fonctions et hauteurs sur les variétés abéli-
            ennes*, Ann. of Math., **82**, 1965, 249–331.

[PSz]   C. Peskine and L. Szpiro, *Dimension projective finie et coho-
            mologie locale*, Publ. Math. IHES, **42**, 1973, 47–119.

[R1]   P. Roberts, *The vanishing of intersection multiplicities of per-
            fect complexes*, Bull. Amer. Math. Soc., **13**, 1985, 127–130.

[R2]   ———, *Intersection theory and the homological conjectures in
            commutative algebra*, Proc. Int. Congress Kyoto (1990), vol. 1,
            361–367.

[R3]   ———, *Multiplicities and Chern Classes in Local Algebra*, Cambridge Univ. Press, Cambridge, 1998.

[R4]   ———, *Recent developments on Serre's multiplicity conjectures: Gabber's proof of the nonnegativity conjecture*, L'Ens. Math., **44**, 1998, 305–324.

[Sa1]   P. Samuel, *Algèbre Locale*, Mém. Sci. Math., **123**, Gauthier-Villars, Paris, 1953.

[Sa2]   ———, *Commutative Algebra (Notes by D. Hertzig)*, Cornell Univ., 1953.

[Sa3]   ———, *La notion de multiplicité en algèbre et en géométrie algébrique*, J. math. pures et app., **30**, 1951, 159–274.

[Sa4]   ———, *Méthodes d'algèbre abstraite en géométrie algébrique*, Ergebn. der Math. **4**, Springer-Verlag, 1955.

[Sc]   G. Scheja, *Über die Bettizahlen lokaler Ringe*, Math. Ann., **155**, 1964, 155–172.

[Se1]   J-P. Serre, *Faisceaux algébriques cohérents*, Ann. of Math., **61**, 1955, 197–278.

[Se2]   ———, *Sur la dimension homologique des anneaux et des modules noethériens*, Symp. Tokyo-Nikko, 1955, 175–189.

[Se3]   ———, *Local Fields*, Springer-Verlag, 1979.

[Sh]   G. Shimura, *Reduction of algebraic varieties with respect to a discrete valuation of the basic field*, Amer. J. Math., **77**, 1955, 134–176.

[ST]   G. C. Shephard and J. A. Todd, *Finite unitary reflection groups*, Can. J. Math., **6**, 1954, 274–304.

[T]   J. Tate, *Homology of noetherian rings and local rings*, Illinois J. Math., **1**, 1957, 14–27.

[W]   A. Weil, *Foundations of Algebraic Geometry*, 2nd ed., Amer. Math. Soc. Coll. Publ., **29**, Providence, 1962.

[Z1]   O. Zariski, *The concept of a simple point of an abstract algebraic variety*, Trans. Amer. Math. Soc., **62**, 1947, 1–52.

[Z2]   ———, *Sur la normalité analytique des variétés normales*, Ann. Inst. Fourier, **2**, 1950, 161–164.

[ZS]   O. Zariski and P. Samuel, *Commutative Algebra*, Van Nostrand, New York, 1958–1960.

Index

Index of Notation